FIRE SERVICE RESCUE PRACTICES

FIFTH EDITION

Cover Photo By:
Elliott Goodman
Washington D.C. Fire Department

VALIDATED BY
THE INTERNATIONAL FIRE SERVICE TRAINING ASSOCIATION

PUBLISHED BY
**FIRE PROTECTION PUBLICATIONS
OKLAHOMA STATE UNIVERSITY**

Dedication

This manual is dedicated to the members of that unselfish organization of men and women who hold devotion to duty above personal risk, who count sincerity of service above personal comfort and convenience, who strive unceasingly to find better ways of protecting the lives, homes and property of their fellow citizens from the ravages of fire and other disasters . . . The firefighters of All Nations.

Dear Firefighter:

The International Fire Service Training Association (IFSTA) is a nonprofit organization that exists for the sole purpose of serving firefighters. IFSTA is a member of the Joint Council of National Fire Organizations, National Fire Protection Association and International Society of Fire Service Instructors. If you need help in locating additional information concerning the organization, training materials, or manual orders, please write or call.

Without you we do not exist

Write:
Customer Services
Fire Protection Publications
IFSTA Headquarters
Oklahoma State University
Stillwater, Oklahoma 74078

Call
(405) 624-5723

First Printing, November 1981
Second Printing, March 1984

Oklahoma State University in Compliance with Title VI of the Civil Rights Act of 1964 and Title IX of the Educational Amendments of 1972 (Higher Education Act) does not discriminate on the basis of race, color, national origin or sex in any of its policies, practices or procedures. This provision includes but is not limited to admissions, employment, financial aid and educational services.

© 1981 by the Board of Regents, Oklahoma State University
All rights reserved
ISBN 0-87939-044-1
Library of Congress 81-82148
Fifth Edition
Printed in the United States of America

CONTENTS

LIST OF TABLES

PREFACE

This edition of *Fire Service Rescue Practices,* the fifth, is a departure from its predecessors. Most important, the subject is being published as a manual by itself, rather than sharing a manual with *Protective Breathing Practices.*

As with all recently revised IFSTA manuals, there are new procedures, new areas covered, more illustrations, and metric equivalents.

Also, this edition of *Fire Service Rescue Practices* covers all the NFPA Standard No. 1001 requirements for certification at Firefighter I, II, and III levels.

To assist the reader in finding information quickly, each chapter divider has a bleed screen at the edge, and an index has been added.

This fifth edition of *Fire Service Rescue Practices* could not have been so extensively revised without the assistance of many people.

Chairman
R. Gene Nichols
Chief, Operations and Training
Springfield Fire Dept.
Springfield, Missouri

Vice-Chairman
Frank Woodruff
Captain, Training Division
Arlington Heights Fire Dept.
Arlington Heights, Illinois

Secretary
Anne Storms
Ann Arbor, Michigan

We also express our thanks to the other loyal and hard-working committee members whose talents and knowledge added to this manual:

Stan Babin
Joseph Beckham
Randy Bennett
Glenn A. Boughton
Owen Collins
James Doyle
Marie Haussermann
Lew Himmelrich
Frank Hubble
Thomas M. Jefferson
Thomas E. Johnson
Donald F. Kehret
Bill Kellum
Mike Litvinchuk
Bradley Malm
Jerry Moore
Carl J. Standefer
Larry Thomas

A special debt of gratitude is owed to Lee Noon, national coordinator of the National Cave Rescue Commission, and Chief Warren E. Isman and Lt. Lawrence Shamer of the Montgomery County, Maryland, Fire/Rescue Services. These three gentlemen gave openly of their time and expertness during the final stages to help round out the manual.

The following generously provided photography, illustrations, and other matter.

Air Safety Cushion, Inc.
Applied Power, Inc.
Construction Safety Associations of Ontario
Nancy Engebretson
Grace Industries
H. K. Porter
Hurst Performance, Inc.
International Association of Dive Rescue Specialists
Journal of Civil Defense
Lisle-Woodridge, IL., Fire District
Louisiana State University
Bill Maggi
Dave Mette
Montgomery County, MD, Fire/Rescue Services
Lee Noon
Paratech, Inc.
Harold Prevost
Elmer Reese
Stoyan Russell
Salem, VA., Fire Dept.
San Rafael, CA., Fire Dept.
Scott Aviation
Shamer and Associates
Dan Smith
Special Service and Supply
Springfield, MO., Fire Dept.
Stillwater, OK., *News-Press*
U.S. Coast Guard
Vetter Systems
Ziamatic Corp.

Tom Gillmore, and Peter Rybacki research technicians, worked long and hard to prepare suitable drafts for the committee. Charles W. Orton brought the validated draft through its final editing stages and into production.

We must also extend our thanks to the following persons whose fine work gave this manual its finished form: Don Davis, production coordinator, Ann Moffat, art, layout, and pasteup; Karen Murphy and Kathy Westerfield, phototypesetters; Carol Smith, proofreading; and Mike Buchholz, copy editor.

Gene P. Carlson
Gene P. Carlson
Editor

THE INTERNATIONAL FIRE SERVICE TRAINING ASSOCIATION

The International Fire Service Training Association is an educational alliance organized to develop training material for the fire service. Each year, validation committees meet at a workshop conference, the objectives of which are to:

develop training material for publication;
validate training material for publication;
check proposed rough drafts for errors;
add new techniques and developments;
delete obsolete and outmoded methods; and
upgrade the fire service through training.

This training association was formed in November 1934, when the Western Actuarial Bureau sponsored a conference in Kansas City, Missouri, to determine how all agencies that were interested in publishing fire service training material could coordinate their efforts. Four states were represented at this conference and it was decided that since the representatives from Oklahoma had done some pioneering in fire training manual development other interested states should join forces with them. This merger made it possible to develop nationally recognized training material broader in scope than material published by any individual state agency. This merger further made possible a reduction in publication costs since it enabled each state to benefit from the economy of relatively large printing orders. These savings would not be possible if each individual state developed and published its own training material.

From the original four states, the adoption list has grown to almost all of the United States; all Canadian provinces; Bermuda; Queensland Australia; the International Civil Aviation Organization Training Centre, Beirut, Lebanon; the Department of National Defence of Canada; the Department of the Army of the United States; the Department of the Navy of the United States; the United States Air Force; the United States Bureau of Indian Affairs; the United States General Services Administration, and the National Aeronautics and Space Administration (NASA).

Every July, IFSTA's validation conference is held in Stillwater, Oklahoma. This conference is an invitational conference, bringing together persons from several related and allied fields:

key fire department executives and drillmasters;
educators from colleges and universities;
representatives of government agencies;
representatives of firefighter organizations; and
engineers from the fire insurance, equipment, and apparatus industries.

The committees on which these persons serve validate the material contained in IFSTA manuals as up-to-date and correct for the fire service.

The publications of the International Fire Service Training Association are compatible with the National Fire Protection Association's Standard No. 1001, *Fire Fighter Professional Qualifications,* and the International Association of Fire Fighters/International Association of Fire Chiefs *National Apprenticeship and Training Standards for the Fire Fighter.* The standards are an effort to attain professional status through progressive training. The NFPA and IAFF/IAFC Standards were prepared in cooperation with the Joint Council of National Fire Service Organizations, of which IFSTA is a member.

To formally adopt IFSTA training material, the firefighters should first request it through their training agencies. Adopting does not obligate any sponsoring agency. It adds prestige to any training program and this, in turn, adds prestige to IFSTA publications.

INTRODUCTION

Fire department rescue functions are wide ranging. Some are directly connected with fires, but many involve anything from massive natural disasters to attempted suicide. Most often the rescue operation is complex and may endanger the firefighter. Fire departments have been the logical anchor organizations to handle most rescue emergencies. The personnel are emergency oriented, trained to remain calm in high-pressure action. Furthermore, much of the equipment used in fire fighting is available for rescue work, the crew members are familiar with operating and maintaining the equipment, and the local fire station is usually strategically located for prompt response.

In this manual, rescue means removing victims from a hazardous or life-threatening situation to areas of safety or treatment. The particular considerations for removing a victim from a variety of hazardous areas will be discussed.

Most rescues have primary and secondary functions:

Rescue has two functions

Primary — Locating and freeing victims and conveying them to a place of safety. Administering first aid as necessary.
Secondary — Restoring the accident area to a safe condition.

There is a close relationship between rescue and first aid practices. Anything violent enough to trap a victim can certainly cause injury at the same time. First aid training for the firefighters or the fire rescue squad is important, and a rescue is not complete unless first aid is handled efficiently. Specific first aid information is available in IFSTA's **Fire Service First Aid Practices.**

Rescuers may have to move a victim without first aid because the situation threatens to deteriorate. First aid treatment under these conditions should be administered as soon as the situation permits.

Rescue functions may need constant consideration

Some rescue operations require constant consideration of primary and secondary functions. For example, a building after a disaster is found to be dangerous to victims, rescuers, and the general public. The slightest movement of debris or wind deviation could cause more debris or structural parts to fall. Fire department rescue workers must always be alert and anticipate such accidents before they happen. It is not easy to juggle the primary and secondary functions of a rescue. Clearing away hazardous debris may require special equipment such as cranes, bulldozers, and trucks to make the area safe again. However, removing of debris after an area is made safe is not a rescue function.

PURPOSE AND SCOPE

All firefighters should be trained in rescue

This manual is an aid to help firefighters expand their proficiency and training in moving victims from a hazardous situation to areas of safety and treatment. Most fire departments have performed rescue work and a community naturally looks to its fire department for leadership during an emergency. Although rescue is often considered a special type of work in the fire service, all firefighters should be trained for this duty. The level of proficiency needed by each firefighter is hard to define, but NFPA Standard No. 1001, *Fire Fighter Professional Qualifications,* offers broad guidelines in the form of behavioral objectives. This manual is designed to meet the behavioral objectives for rescue skills for Fire Fighter I, II, III.

NFPA 1001 specifies, in terms of performance objectives, the minimum requirements of professional competence required for service as a firefighter. In many instances, this manual illustrates just one method for accomplishing each task. The methods shown throughout the text have been approved by the International Fire Service Training Association as accepted methods for accomplishing each task. However, they are *not* the only authorized methods. Students should use the methods specified by the authority having jurisdiction.

This rescue manual is divided into topical chapters. At the beginning of each chapter, those behavioral objectives covered in that chapter are outlined as they appear in NFPA 1001. Where feasible, the content of each chapter reflects this order. The standard does not require objectives to be mastered in order. The local or state training program should establish instructional priority and training program content to prepare individuals to meet the performance objectives set forth by NFPA 1001.

Mental And Emotional Crises Of Rescue

Chapter 1

Chapter 1

Mental And Emotional Crises Of Rescue

An accident may cause many normal reactions, but some are severe enough to require aid. Every person will be stunned and temporarily disorganized, but most will probably adapt themselves quickly and respond positively to the demands of the situation.

Frequently, victims will require more than rescue because they may be so emotionally overwhelmed that their reactions may become exaggerated and dangerous to themselves or others. Family anxiety is natural and time given to needed support may prevent persons from collapsing from shock, family members from injuring themselves, or hysteria from spreading to crowds.

Rescue situations affect victims and rescuers

Rescuers bring their own special reactions, values, and capabilities with them. They will always be affected by the stressful situations under which they work, but their reactions must not affect their ability to perform.

THE VICTIM

Response to an accident or trauma may be physical and/or emotional.

A *physical response* is a person's unconscious method of coping with the immediate situation. *Great care and caution must be taken (especially by an untrained person) in concluding that no physical impairment is present.*

Emotional shock. Victims with no physical injury may experience an emotional shock, exhibited by rapid respiration and heart beat, profuse perspiration, trembling, and weakness. More severe reactions are nausea and vomiting. Emotional shock should be cared for in the same way as physical shock.

Conversion hysteria. Victims may believe incorrectly that some part of the body has ceased to function. Without physical basis, they may believe that legs or arms will not work. Do not consider them malingerers and do not try to convince them that their legs or arms do work. If time permits, the rescuer may splint or treat the part to soothe the victim.

Migraine headaches. These may be caused by emotional reactions that make certain blood vessels in the brain constrict. Frequently, the headache starts on one side of the head and causes nausea and dizziness. The victim should rest quietly.

Response to situation may be physical

Cardiac neurosis. A normal heart may react to stress and the victim may experience reactions resembling a heart attack. Pain in the left side of the chest, dizziness, and fainting may occur. These symptoms increase the victim's fear.

Hyperventilation. Caused by anxiety, hyperventilation is expressed by rapid breathing that results in overoxygenation and carbon dioxide depletion in the blood stream. The victim may faint or stop breathing if the condition continues.

Emotional response to a traumatic situation may be expressed in any combination of the following conditions:

State of disorganization. A person is not yet able to decide the appropriate actions to take in a stressful situation, but could follow the actions of a loud panic-prone leader and create further disorganization.

State of depression. A person in this state could be numb and incapable of action. Such a person is genuinely unable to assist or help in the situation and should therefore be treated calmly and gently rather than with hostility.

Response to situation may be emotional

Panic-prone reaction. This reaction is expressed by hyperactivity and extreme jitters and may turn into hysteria. A hysterical person may initiate mass panic. To deal with the situation effectively, the rescuer needs to display a calm, composed manner. If restraint of the person becomes necessary, the rescuer needs ample help to control the person quickly and calm any crowd present. The rescuer must not slap or shake the victim since this may increase the victim's and the crowd's anxiety.

Paranoid reaction. Paranoid reaction, an abnormal result of frustration, is expressed by an angry, uncooperative, and overly suspicious attitude. A person exhibiting such actions usually misunderstands attempts to help. How to handle such a victim will depend on the basis of the paranoia. There are two general categories of paranoia. First, there is the mentally unstable individual who because of some emotional and traumatic experience reacts abnormally to assistance; second is the paranoid drug user who because of drug-induced psychosis reacts abnormally to rescuers. The rescuer should first make sure the victim has no weapons available and that there is sufficient help. Often, police are a good source of help; however, in some instances the presence of police increases the victim's anxiety, especially in drug-induced paranoia. All actions must be explained to the victim in a calm voice. The victim should be reminded that a firefighter is attempting to help. The rescue should be done in a slow, calm, diplomatic, but forceful manner. If it becomes absolutely necessary to overpower the victim, ample assistance and restraints must be available.

No one thing will work regularly to help even a majority of victims, because their individual reactions to the stressful accident will vary according to their individual backgrounds. The main thing for the rescuer to remember is to care. A professional attitude on the part of the rescuer is desirable, but this does not require the rescuer to be cold, callous, or dispassionate. For the rescue worker, the best combination of personal traits is compassion and professional expertness.

Be professional, and care

In addition to establishing a more personal contact or rapport with a victim, rescuers can improve the victim's emotional state by providing as much information as possible. Some wrongly feel that the truth may cause anxiety. Victims and family members without complete information or without the experience to assess what they see, often have fearful expectations or fantasies much worse than the real accident.

Some guidelines for informing victims are listed below:

- Gently try to orient the victim to the surroundings, addressing the victim by name, if possible, and explaining that there has been an accident.
- Explain who you are and what actions are being taken on the victim's behalf.
- Avoid glum comments about physical condition, even to those nearby. Remember that semiconscious victims have reported overhearing such remarks.
- Assure the victim that you will locate and notify relatives.
- Be honest without shocking the victim. Describe the scene only generally.
- Allow for hope. Do not tell a victim that death is near unless directly asked. Then, do not lie. Use discretion in relaying news to the victim about the death of a family member or others.

Use discretion when informing victim

Rescue workers have two responsibilities under the primary function of rescue:

- To free trapped victims and move them to safety.
- To cushion the potentially harmful effect of the stressful circumstance, thus preventing an emotional crisis that could lead to physical harm at the scene or later.

THE FAMILY

The family also deserves attention

From a rescue standpoint, it may not be a high priority to deal at length with a victim's family, but the family certainly deserves considerate attention to lessen their anxieties.

If possible, let the family see the victim. Even if the victim is dying, consider allowing the family to remain with the victim. Clinical experience shows that making the family leave only adds to their grief.

Family members may react with anger and even express verbal aggression toward the rescuer, who must assume an impersonal attitude.

THE RESCUER

Rescue workers should not participate in situations with which they cannot deal. Such situations might include rescuing family members, dealing with severe burn victims, or handling other catastrophic emergencies. The rescuer should report to the officer in charge and request to be relieved as soon as the situation permits the rescuer to leave.

The officer in charge of a rescue scene should be aware of the psychological condition of the rescue workers. If the officer in charge notices a rescuer who is severely affected by the situation and not functioning properly, the officer in charge should remove that person from the scene as quickly and quietly as possible without letting the rescuer lose respect or dignity because of the inability to cope.

This action on the part of the rescuer or the chief officer should greatly increase the atmosphere of control in a rescue and help the affected rescuer minimize potentially long-lasting psychological effects.

The rescuer is involved

Far from being an uninvolved individual helping in a rescue, the rescuer is part of the situation. No matter how professional and detached the rescuer attempts to be, emotions get involved. This is natural, and probably is one of the reasons rescuers can be effective. Through this empathy and personal involvement, the rescuer can respond to the victim's emotional needs. However, this involvement can develop into emotional problems for the rescue worker.

It is generally impossible for the rescue worker to turn off this involvement or forget about it when the emergency is over. Depending on the rescuer's ability to cope with this involvement and the duration and magnitude of the situation, the rescue worker may need help adjusting emotionally. Other emotional problems may follow the emergency. The most important thing to remember is that it is natural, and that asking for help is not a sign of weakness.

Rescue workers who are called to catastrophic emergencies seem to be more likely to need assistance. Rescue workers who respond to a high incidence of calls may become demoralized or develop guilt feelings and need counseling. The hardest part is recognizing the need for help.

Rescuers may need help adjusting

There are several signs that indicate a rescue worker may be having trouble adjusting, and they can appear in any combination or order. The following are some of the more common:

- *Prolonged disorientation or lack of action at emergency scene.* Initial or momentary disorientation is natural, but if it lasts several minutes or interferes with the individual's normal functions, it may indicate adjustment problems.
- *Prolonged or recurring depression.* Certainly depression or being upset is natural, but if it lasts for a long time or interferes with the job or family life, the individual should seek assistance.
- *Uncontrolled crying.* Crying is a normal and healthy outlet, but if the crying does not stop after a reasonable period of time, or if it recurs for a long period after the incident, it may be considered an indication of maladjustment and a need for help.
- *Nightmares or inability to sleep.* Nightmares can be expected to last several weeks following extreme instances, and trouble sleeping for several days, but if the condition becomes chronic or habitual, help should be sought.
- *Lack of or elevated appetite.* It may be hard to eat following a gruesome emergency. This lack of appetite again must be measured by the length of time it lasts. More than a couple of weeks indicates difficulty. Some persons may react by overeating. This reaction may not be so obvious, but a weight gain of more than 10 pounds (4.5 kg) in six weeks could be suspect.
- *Irritability and outbursts of anger.* Like most other emotional reactions, irritability and anger are normal reactions to stress and frustration. Unfortunately, they do not

enhance emergency operations, and if the outbursts become chronic or interfere with the individual's ability to function, help should be sought.

- *Alcohol or drug abuse.* Of all the symptoms of poor adjustment, alcohol or drug abuse can be the hardest to detect and treat. Generally, professional help is needed to overcome this maladjustment.

These signs may be noticed in fellow workers or they may be noticed in oneself. However, when they are noticed there is no need to be fearful or ashamed. What if a firefighter seems to be having problems—what can be done? The basic approach is to talk about the situation to co-workers or close friends, doctors, ministers, or psychologists. Feelings should be discussed and explored. If someone else is having trouble adjusting, encourage the person to talk by listening, or suggest a talk with someone else, or inform supervisors so they might help.

Other ways to help are listed below:

- Realize these feelings are a normal reaction to stressful work.
- Those with religious convictions may find praying advantageous.
- Help others cope with their problems. This can provide additional insight into individual personal problems.

ASSISTING AND TRANSPORTING VICTIMS

When victims are in a dangerous area they must be moved as quickly as possible. The methods are many and varied. Firefighters should be trained and efficient in all moving techniques, and then choose those carries or drags with which they feel most comfortable.

Stabilize injuries

A victim should be examined for injuries and stabilized before moving, if at all possible. A victim with a fractured extremity should have the extremity immobilized before being moved. Moving an injured victim without first stabilizing the injuries can lead to further injury or aggravate the condition of the victim. Every effort should be made first to determine the injuries of the victim, stabilize the condition, and then move the victim. Unfortunately, this ideal cannot always be realized. A victim with a fractured arm may be in a toxic atmosphere. The first requirement would be to get the victim to a safer atmosphere and then stabilize the fractured arm. A rescuer many times will have to use personal judgment in moving victims. If it is apparent that a victim is injured, every reasonable effort should be made to protect the injured part during movement, regardless of the mode used.

Rescue Tools

Chapter 2

NFPA STANDARD 1001
FIRE FIGHTER II

4-12 Rescue

4-12.2 The fire fighter shall demonstrate the use of the following rescue tools:

(a) Shoring blocks

(b) Trench jacks

(c) Block and tackle

(d) Hydraulic jacks

(e) Screw jacks

Reprinted by permission from NFPA Standard No. 1001, *Standard for Fire Fighter Professional Qualifications.* Copyright © 1981, National Fire Protection Association, Boston, MA.

Chapter 2
Rescue Tools

The skills and techniques required for the complexities of rescue work can be gained only through a sound program of training. Because there are so many possible rescue situations involving from one victim to perhaps hundreds, it is impossible to train for every contingency. Rescue personnel can best be ready to handle any situation if they are proficient with their rescue tools. Complete knowledge of tools and equipment allows rescuers to rapidly devise a method for the rescue at hand.

Rescuers need continual training

Basic rescue tools are carried on all fire apparatus. Different rescuers, planning for a wide range of emergencies, would provide different lists of all the other necessary equipment a rescue team should have. The actual decision on which equipment to purchase can only be made locally. Where the capability exists, the use of good homemade tools should not be overlooked.

Tool lists are determined by the type and size of rescue unit in service or planned by a rescue team. Then the list is modified according to the probable rescue needs of an area. If rescue planners consider their real needs, they can probably avoid purchasing an unnecessarily large vehicle at great expense, poorly planned use of storage space, and early obsolescence or costly modification.

With a particular type of rescue vehicle in mind, or even if all rescue equipment is to be carried by an engine or truck company, consideration must be given to the local environment and circumstances. Rescue tools would differ if the rescue vehicle or engine serves a rural area or a heavily traveled highway. How far is the area from a hospital? Longer distances might suggest more first aid and life support supplies. Chemical plants, oil refineries, and many other occupancies will require combustible-gas detectors.

Choose tools according to local needs

The following is not a checklist of all needed equipment. It may be used best as a general guide for possible choices. Personnel should be completely familiar with all tools selected for local use, both to use them efficiently and safely and to maintain them in good working order. Protective clothing should be worn when using all tools.

STRIKING TOOLS

Some of the most common and crude rescue tools are striking tools (Figure 2.1), most of which are characterized by large, weighted heads on handles. This category of tools includes axes, battering rams, ram bars, punches, hammers, sledge hammers or mauls, chisels, automatic center punch, and picks.

Figure 2.1 Examples of striking tools for rescue.

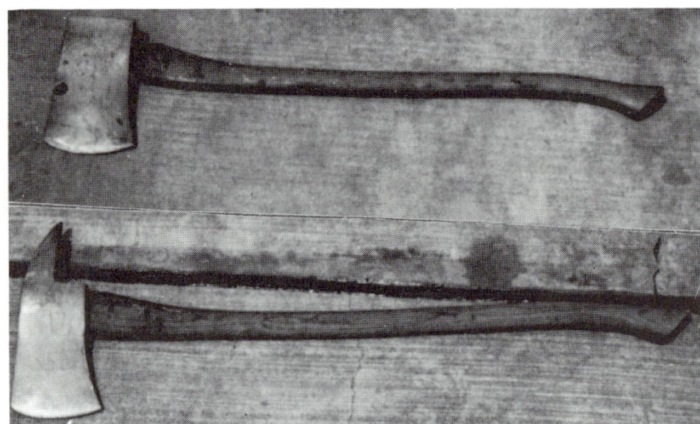

Axes

Battering ram

Hammers *Courtesy of Elmer Reese*

Sledge hammer *Courtesy of Montgomery County Fire/Rescue Service*

Firefighters should seriously consider using hand tools such as these before using power equipment. A victim can often be rescued with hand tools in the time needed to get the power equipment ready for use. A flathead axe and a hammer used together, for example, will quickly cut most sheet metal found in automobiles. The ram bar can be driven into the lock area of an older auto not equipped with safety door locks to break the lock.

Use striking tools with care

Striking tools are dangerous. They can crush or sever fingers, feet, or any other part of the body. They can also drive chips and splinters into the air, piercing skin and eyes. Proper protective clothing must be worn: boots, turnout coat and pants (if available), gloves, goggles, and helmet. All tools should be properly maintained. The handles should be solid and well set in the head. The striking surface should be serviceable: axes or pointed tools should be sharpened; blunt striking surfaces should be free of chips or cracks. Striking tools should be used with short, quick strokes. Long, sweeping strokes are uncontrollable.

PRYING TOOLS

Prying tools (Figure 2.2) are also common in the fire service. Many manufacturers have designed tools especially for fire service rescue jobs. The Pry-Axe, Halligan, claw tool, pry bar, Kelly tool, and Quic-Bar all share the same operating characteristics. Prying tools are used to open doors, windows, hoods, and trunk lids of automobiles; to open the doors or to break the locks in structures; and to lift automobiles and more heavy objects. Crowbars and other prying tools are excellent for starting or widening an opening for larger power tools.

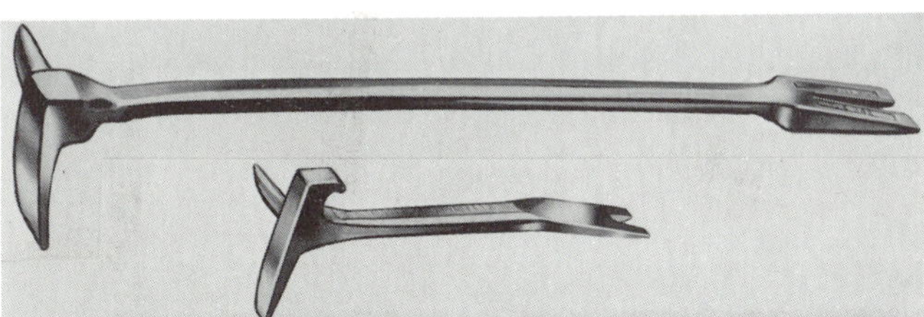

Quick-Bar & Mini Quick-Bar

Courtesy of Ziamatic Corp.

Figure 2.2 Examples of prying tools for rescue.

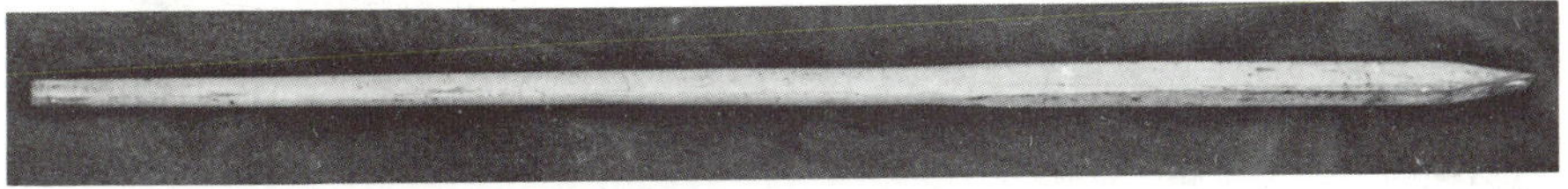

Pry Bar

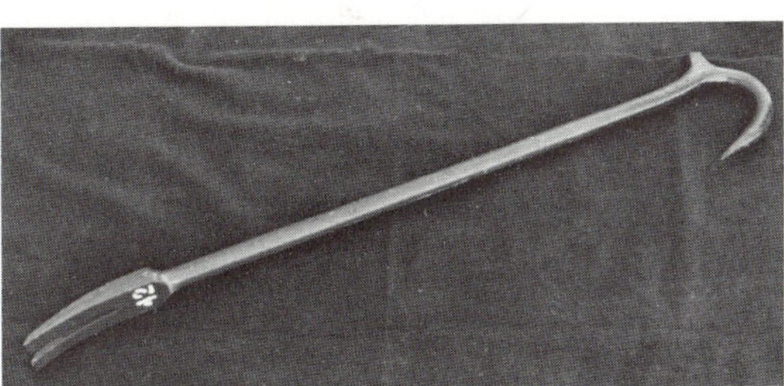

Claw tool

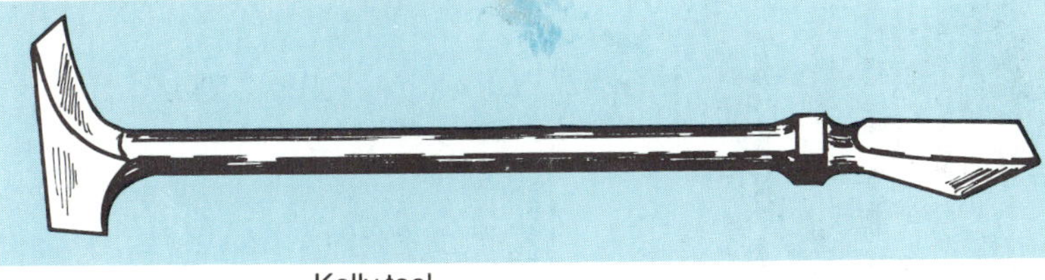

Kelly tool

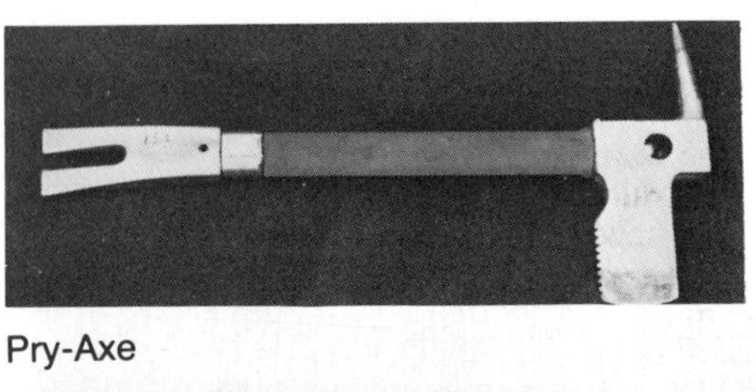

Pry-Axe

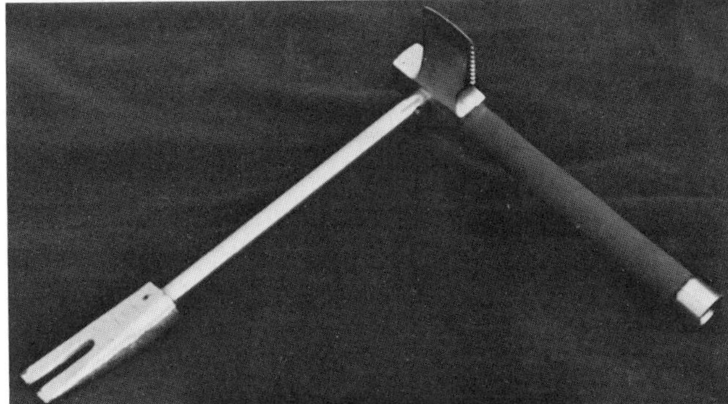

Pry-Axe

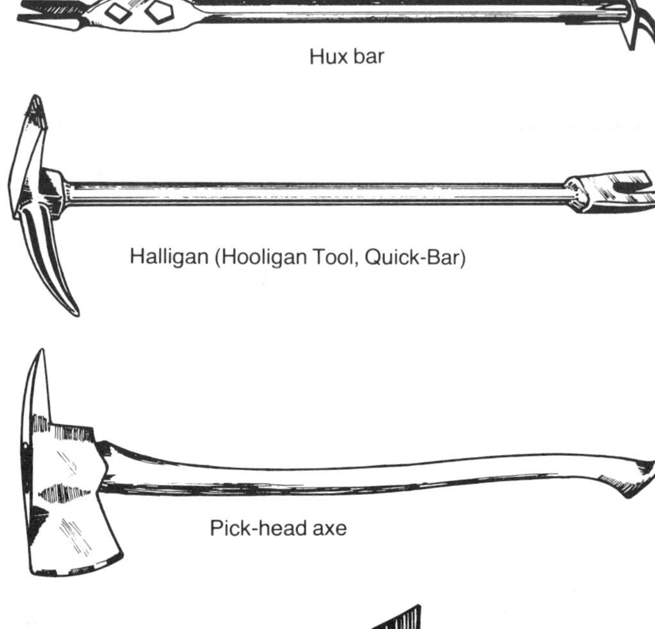

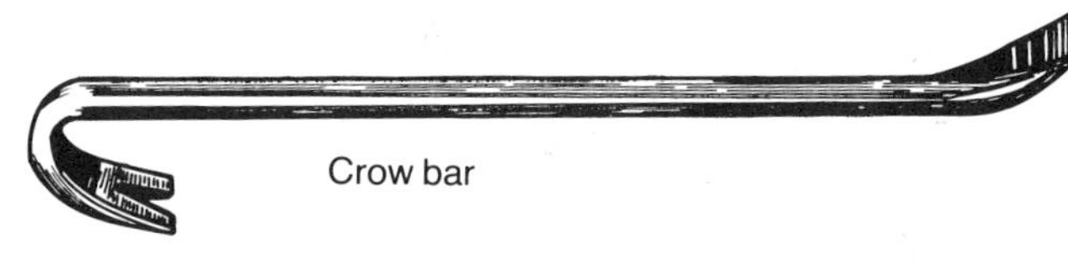

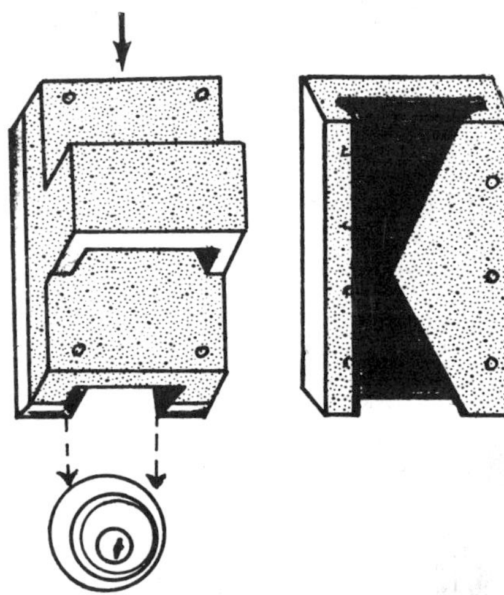

Figure 2.3 A K-tool is good for pulling the lock from a door.

Stopped or
Rabbeted Jamb
Casing
Door
Lock Bolt

Figure 2.4 Doors may be pried until the bolt slips the keeper.

Several door openers operate using the leverage principle. Doors are pried open in three basic ways: by breaking the latching mechanism, by breaking the keeper out of the jamb, or by separating the door out of the jamb far enough to slide the bolt out of the keeper (Figures 2.3 and 2.4).

Prying tools are generally more safe than striking tools when used correctly. The worst safety hazard associated with prying tools is the temptation to use them incorrectly. It is not acceptable to use a "cheater bar" or to strike the handle of a pry bar with other tools. This can seriously injure a firefighter and often will ruin the tool. If a job cannot be done with one tool, use another. Do not use a prying tool as a striking tool unless it has been specifically adapted for that use.

CUTTING TOOLS

Cutting tools use a wide range of power sources. Most of the tools are characterized by a sharp, hard metal cutting edge, but others have an abrasive cutting edge. The primary and distinguishable difference between a cutting tool and a striking tool with a cutting edge is that the cutting tool is designed to produce a precise and controlled cut, whereas a striking tool provides a more crude, jagged cut. Cutting tools include knives, saws, chisels, torches, wire cutters, bolt cutters, powerline cutters, tin snips, boring tools, and scissors (Figure 2.5).

Figure 2.5 Examples of cutting tools for rescue.

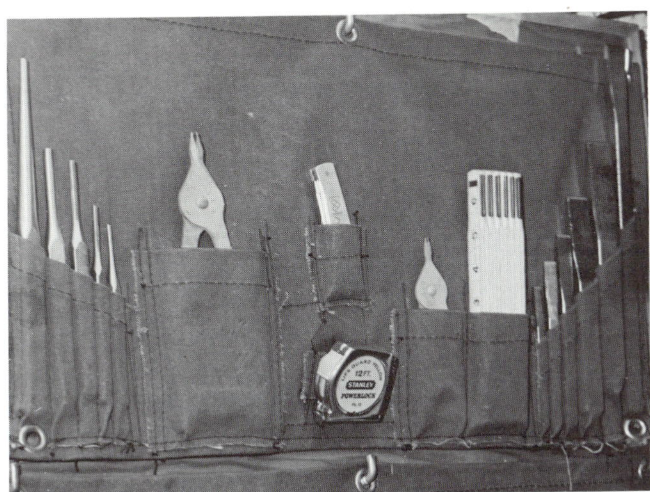

Chisels, metal cutters, wire cutters, and pliers
Courtesy of Elmer Reese

Hydraulic circular saw

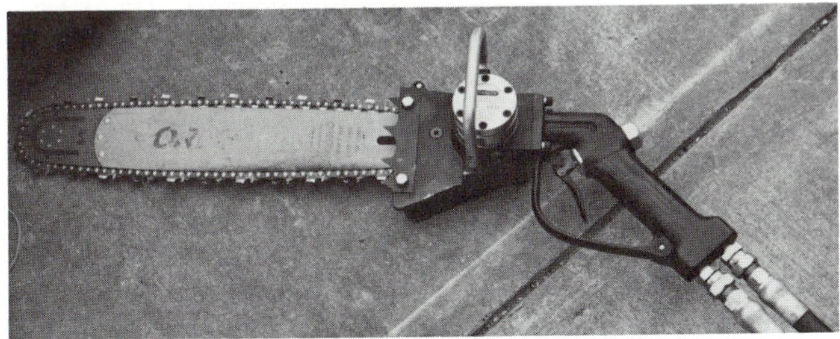
Hydraulic chain saw

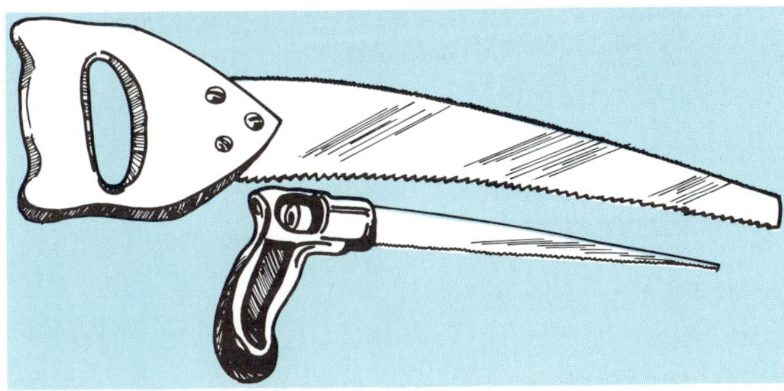
Saws

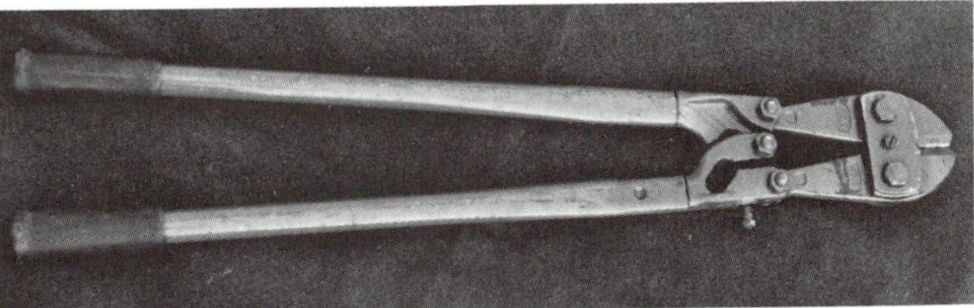
Bolt cutter

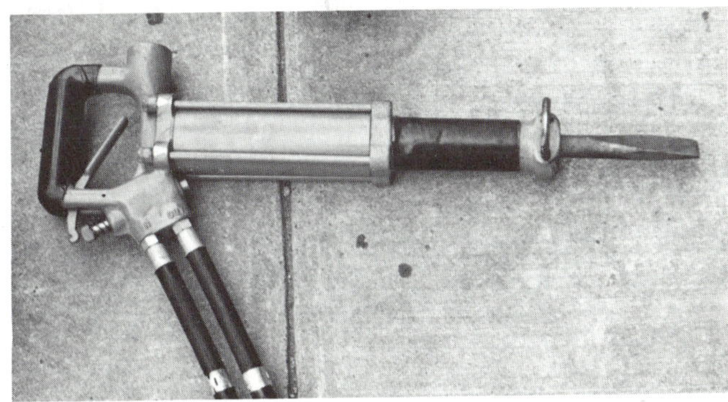
Hydraulic chisel

Oxyacetylene torch

Engine-driven chain saw

Two-man saw

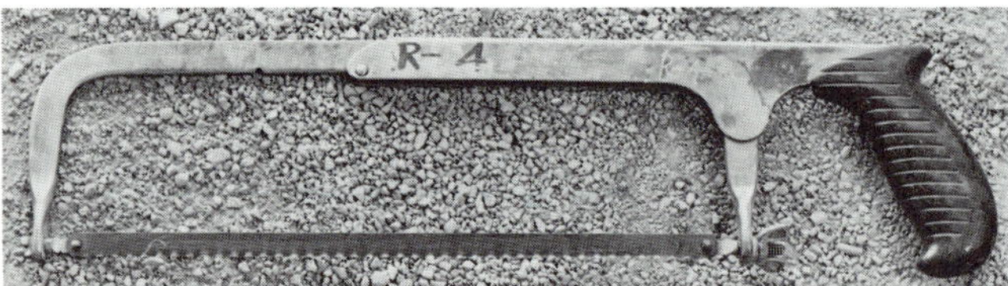

Hacksaw

Engine-driven circular saw *Courtesy of Paratech, Inc.*

Axes can be used to remove glass and to cut metal. Pickhead axes can be used to help hold come-alongs in place so units do not slide on hoods, fenders, and trunks. Hatchets are excellent glass removal tools, easier than axes to control and able to transmit the force of a six-pound (3 kg) axe.

Cutting tools often misused

Cutting tools are the most diversified of the tool groups, which leads to much misuse. Cutting tools are designed to cut only a specific type of material. The misuse occurs when a tool is used to cut material it was not designed to cut. This misuse can destroy the tool and endanger the operator.

Cutting tools commonly misused are the opposing-jaw metal cutters. These include wire cutters, bolt cutters, and energized wire cutters. These tools may look alike, but their use is not interchangeable. The most dangerous misuse is to attempt to use bolt cutters for energized-wire cutters. This can result in electrocution. Only cutting tools approved by a recognized agency such as Underwriters Laboratory or Factory Mutual should be used to cut energized electrical wires, and then only in a life-threatening situation when no other means of disconnecting the electrical current is readily available. These tools must be maintained in strict accordance with the manufacturer's recommendations and used only when the rescuer is wearing approved lineman's gloves. Energized-electric-wire cutting tools should be clearly marked and maintained for their intended purpose, and used for no other purpose.

Rotary-blade saws are good for building rescue, ventilation, and door and wall cutting. Rotary-blade saws come with interchangeable blades, and each has a specific cutting capability. If the blades are used to cut materials that they are not intended to cut, they may fracture and spin apart with explosive power. Time must be taken to change to the proper blade. A rotary-blade saw or any power saw should only be guided by the operator. The saw should be allowed to do the work. Forcing the saw only damages the saw and blade, possibly causing the blade to fracture.

Rotary-blade saws dangerous around autos

If at all possible, they should not be used for vehicle extrication because of their apparent fire hazard. If the saw is all that is available, use it with extreme care. Cover all victims and rescue personnel in or near the vehicle, and have at least one charged line standing by. The type of vehicle and hazard and the placement of the vehicle will dictate whether more than one line will be needed. While using the saw, cool the blade with a fine water mist from a booster line. *Caution*: Make sure the water mist is on the blade before cutting any metal; blade disintegration may result from putting water on a hot blade. Wear full protective clothing with face and eye protection because rotary-blade saws discharge molten metal. Also be careful not to use one manufacturer's blade on another's saw. Blades from different manufacturers may look alike, but they should not be used on tools from other manufacturers. Always store blades in a clean, dry environment.

The oxyacetylene cutting torch is a versatile cutting tool that can be used to gain entry quickly through metal barriers. It has specific application in cramped and tight locations or for cutting materials under slight tension that would tend to bind a saw blade. Another excellent use is to remove flush or rounded bolt heads. If at all possible, do not use cutting torches in vehicular extrication because of the extreme fire hazard.

The cutting torch actually burns the metal in an oxygen-enriched atmosphere. The torch preheats the metal to ignition temperature and then the oxygen handle is depressed to provide pure oxygen for the metal to burn. But the potential source of ignition can prove extremely hazardous. In addition, acetylene has a wide explosive range and accumulation of this gas in closed spaces can result in violent explosions.

Watch for fires started by cutting torch

Anyone using a torch should be properly equipped with welder's goggles and gloves. A second person should stand by to watch for fires that the torch could start. If the cutting is to be done around flammables, a suitable portable extinguisher and/or a charged hoseline should be handy. If necessary, wet down the area before the cutting begins.

The user has to be fully trained and sure that the cutting equipment is properly set up. The threads on cutting equipment are opposite, so it is virtually impossible to put the hoses on incorrectly. The hoses are color coded: green for oxygen and red for acetylene (Figure 2.6). The three major concerns of the operator are that the regulators are properly adjusted, the cylinders are secure, and the cylinders are upright. Excessive pressures may mix the gases inside the equipment or decompose the acetylene in its hose, causing an explosion. The manufacturer's instructions should be consulted for proper pressure settings. If the cylinders are not secure and fall, the valve stem may break off. This may turn the cylinder into a rocket or cause a large pocket of explosive gas. Acetylene cylinders cannot be safely used in any position except upright. Generally, the torch will not function properly if the cylinders are not upright (Figure 2.7). In any other position a loss of acetone, an acetylene stabilizer, may occur. Because of the unstable nature of acetylene, this loss may result in an acetylene explosion within the cylinder.

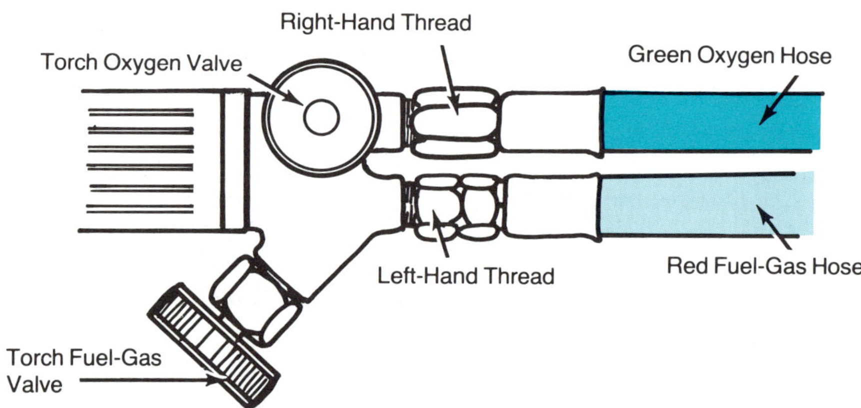

Figure 2.6 The hoses and connections of an oxyacetylene torch are coded so the user cannot make disastrous mistakes.

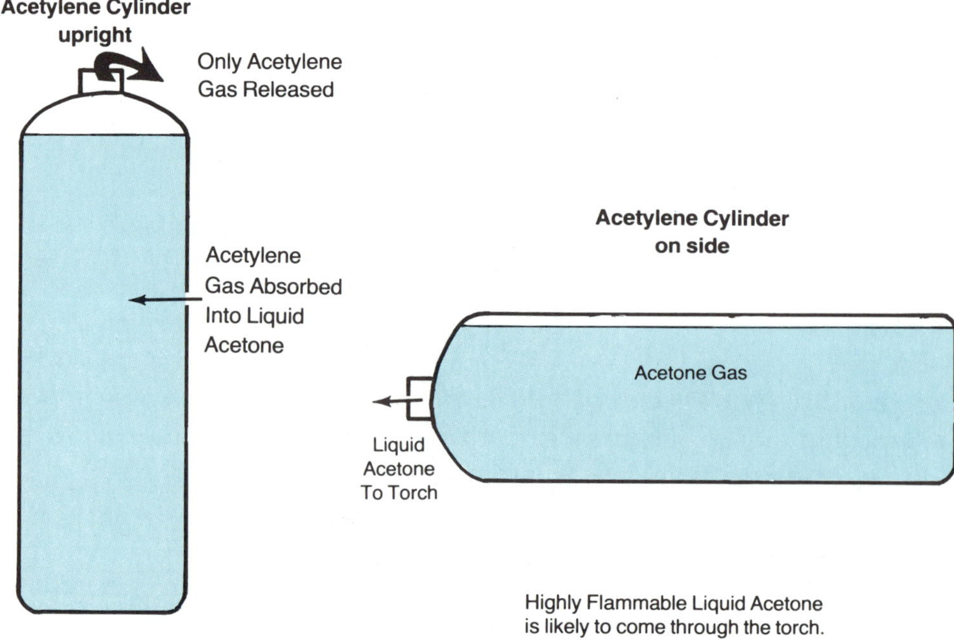

Figure 2.7 When an acetylene cylinder is on its side there is danger of liquid acetone coming through the torch.

The group of cutting tools used for boring or drilling includes hand augers, brace and bits, and electric and hydraulic drills (Figure 2.8). These tools can be used to start saw cuts, open locks, and drill out bolts, rivets, or screws for access. The hazards of the tools are their sharp points and the rotary action of the bits.

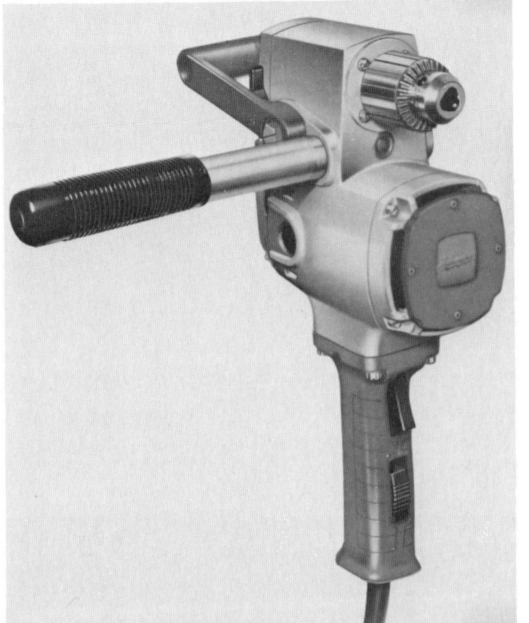

Electric drill *Courtesy of Milwaukee Electric Tool Corp.*

Figure 2.8 Examples of boring tools for rescue.

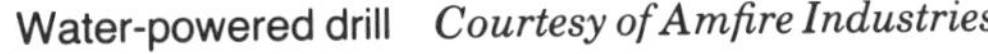

Water-powered drill *Courtesy of Amfire Industries*

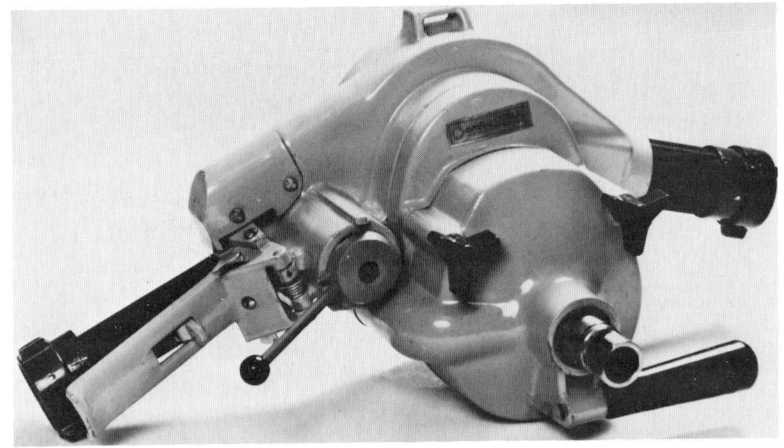

Chisels are helpful and versatile. However, if used improperly they are dangerous. Most chisels are used with a striking tool, and each is designed for a special purpose. The head and metal vary to meet different needs. A chisel must be examined for defects before use. A chisel with a blunt tip or mushroomed head should not be used. The chisel should be driven with a light tapping instead of hard blows. If hard blows are necessary, the chisel is not the right tool to use.

Pneumatic-powered chisels (Figure 2.9) are useful for rescue work. The chisel can be powered by a breathing apparatus air

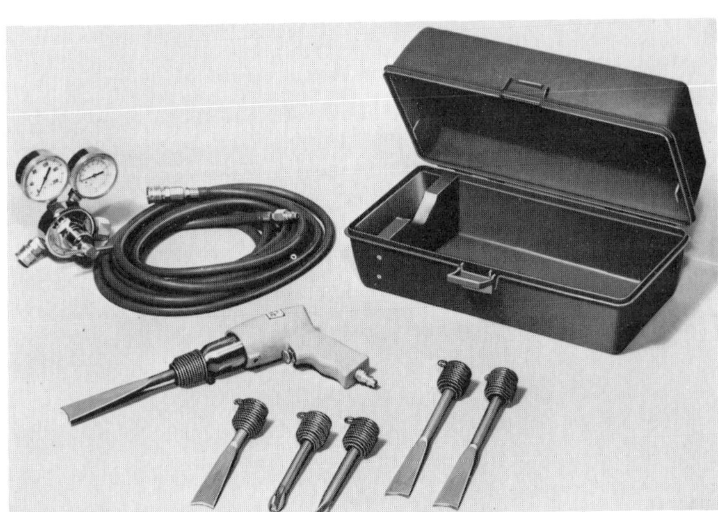

Figure 2.9 Pneumatic chisels may be powered by compressed breathing air, carbon dioxide, or nitrogen. *Courtesy of Ziamatic Corp.*

cylinder, a cascade system, or a truck's airbrake system. It can also be adapted for use with cylinders of carbon dioxide or nitrogen. The air chisel can be especially effective for auto extrication, cutting through the top, roof support columns, seat bolts, and door lock assemblies. It is excellent for cutting medium- to heavy-gauge sheet metal and for popping rivets and bolts.

LIFTING OR PULLING TOOLS

Often, the firefighter has to lift or pull an object or victim. The object may weigh several hundred pounds, and the lift or pull may range from only one or two inches to several feet. Several rescue tools have been employed to assist in this task.

The most common lifting tool is probably the jack. There are three basic types of jacks: the screw, the ratchet lever, and the hydraulic (Figure 2.10). The application of each is similar, but the operation is different.

Figure 2.10 The three types of jacks useful for rescue are the hydraulic, ratchet, and screw.

The simplest is the screw jack. There are two types of screw jacks: the folding and the bar. Both jacks have a male-threaded core similar to a bolt and a means to turn this core. Also, both have a female-threaded base to match the core. Turning the core moves it through the base in a screw action on the bar type. The core extends from the base, lifting as it is turned. The action of the folding jack is somewhat more complicated. The frame of the jack is made of metal bars of equal length fastened together in the center to form Xs. These Xs are then fastened together top to bottom, two or three high. Typically, two sets are set side by side for stability. The bottom ends of the bottom X are attached to a base that allows them to move together or apart along a track. The top of the bottom X is attached outside to a female-threaded nut similar to the one in the base of the bar type. The other is attached to a bushing that is not threaded. A male-threaded core with a head is put through the bushing and threaded into the nut. As the core is turned, the bushing and nut come together, causing the jack to extend and lift.

The ratchet-lever jack uses the lever principle to do work. The jack is a solid metal bar with teeth up one side. The jacking carriage fits around the bar and is double ratcheted on the geared side. One ratchet holds the carriage underneath. The second ratchet is combined with a lever that is pushed down to force the carriage upward. The ratchets can be reversed to move the carriage down.

The hydraulic jack operates on the principle that the pressure of liquids between two chambers of unequal size tends to equalize. A small chamber is used to pump fluid into a larger chamber. The energy applied to the small chamber is multiplied by the surface area differential to do more work in the large chamber. There is a check valve between the chambers that keeps the liquid from flowing back.

Each jack has best use

These three jacks are all used in rescue work, but each is better for certain tasks. The bar screw jack is an excellent tool for holding loads under compression, such as in shoring or similar work. The power magnification of screw jacks is limited, which tends to make lifting difficult. Bar jacks have also been used effectively as hangers for smoke ejectors. This has been accomplished by setting them across the top of a doorway or window frame, applying enough compression to hold them in position, and then hanging smoke ejectors or salvage covers from them.

The folding jack is of limited use. It is considered safe only for light loads. The major advantage of this jack is that it can be collapsed down to fit in tight places (four- to six-inch [10 to 15 cm] clearance).

The ratchet jack can be a good medium-duty jack, but it is by far the most dangerous. Attaching the carriage to the object being lifted is often difficult because of the height of the lifting stand. The ratchets have also been known to fail under a heavy load.

The hydraulic jack is the best heavy-duty jack for most situations. Its lifting capacity and dependability have made it popular. In addition to being a good lifting jack, the hydraulic jack is an excellent compression jack for shoring operations.

Lifting any object is hazardous and lifting with a jack is no exception. The primary consideration in jacking is good footing. Remember that the load will be transmitted to the base of the jack. The jacking surface should be solid, flat, and level. If it is not, a flat board or steel plate should be put under the jack and shimmed level. Never work under a load being held unassisted by a jack. The load must be shimmed or cribbed to make sure it will not fall.

Another lifting tool often used is the come-along, which increases the pulling ability of a lever to its maximum through ratchet action (Figure 2.11). Typically, the come-along has a drum that is rotated by the handle directly or through gear reduction. A cable or chain is attached to the drum, which makes the effective pull of the come-along equal to the length of the cable. The come-along is typically anchored to a secure object and the cable run out to the object to be moved.

Figure 2.11 One type of come-along.

Special-service winches mounted on vehicles and tow trucks work even better than the come-along. Their use as lifting tools may require special skills, or additional heavy rescue equipment such as rigging. Do not use ladders as rigging with winches or tow trucks, because the power involved could easily destroy a ladder.

The air bag (Figures 2.12 and 2.13) is easily inflated by an SCBA air cylinder, and is an especially useful lifting and pulling tool. Made of puncture- and cut-resistant material, it can lift very heavy objects, help to stabilize wreckage, or pull a steering column out of the way.

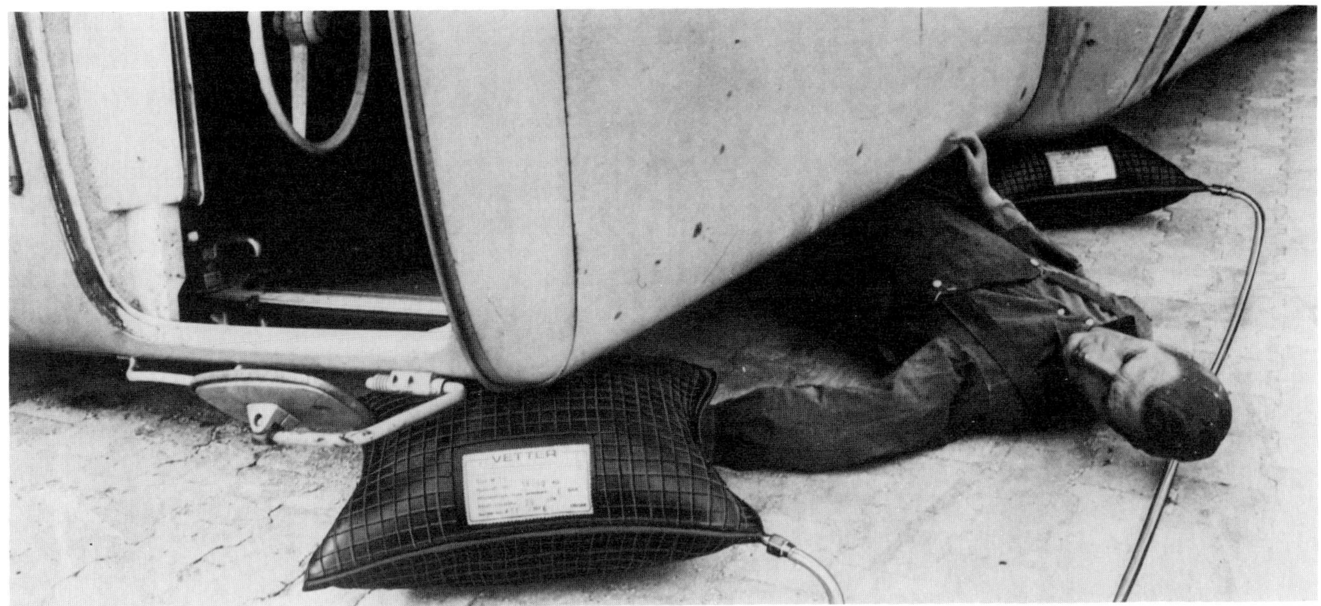

Figure 2.12 Simulated rescue, illustrating the air bag's utility in freeing pinned victims. *Courtesy of Vetter Systems.*

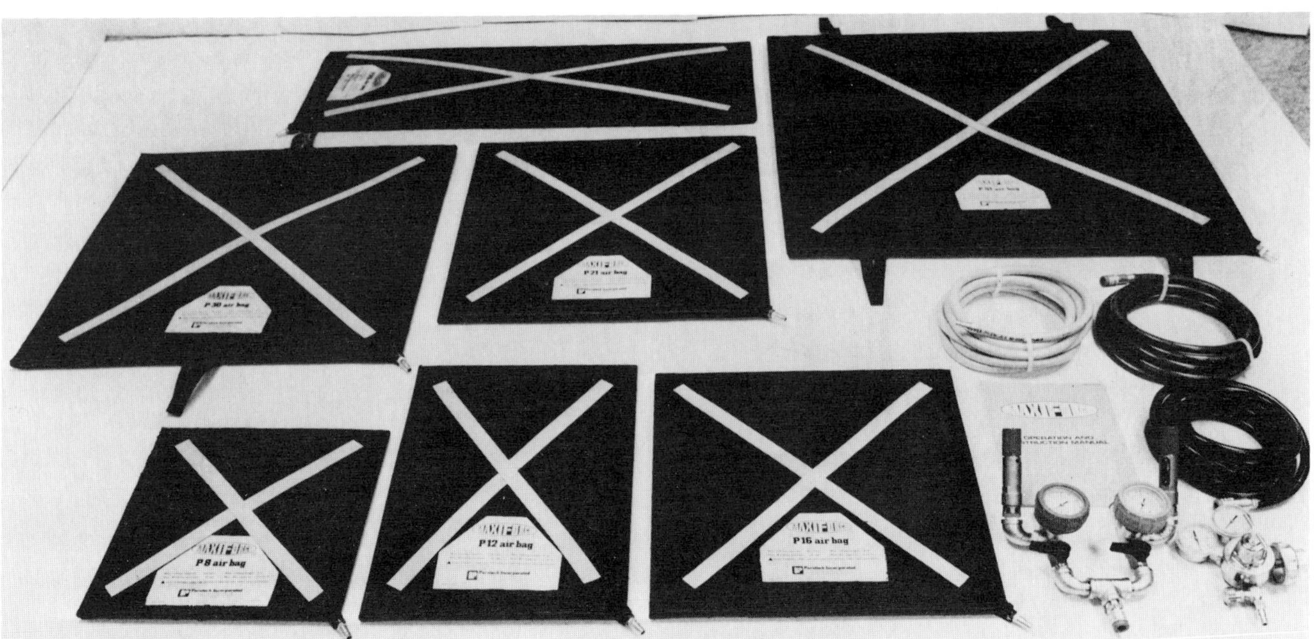

Figure 2.13 Air bags are manufactured in assorted sizes. When deflated they are slim and easily handled. *Courtesy of Paratech, Inc.*

Shoring blocks, ringstands, or other shimming devices must be used with the jack in rescue lifting operations. Shoring blocks can be made beforehand out of different lengths and widths of hardwood and carried on rescue vehicles. Generally they are preferred over ringstands because they are less expensive and require less compartment space.

Some departments have color coded their shoring blocks to indicate length and width. The length is represented by the color the whole block is painted; the width by a colored stripe. If the block is wedge shaped, it has no stripe. This allows quick reference on the emergency scene while the blocks are still in the compartment. A tape measure is not needed—the block size is known when the blocks are pulled from the rescue vehicle compartment. Loops of webbing cord or rope may be attached to the end of shoring blocks to remove them easily (Figure 2.14). When possible, shoring blocks should be shoved into place with a second block or a short tool. This avoids having hands under the load.

Ringstands are popular in some instances because they are adjustable and stable. If shoring materials are not available, materials at hand can be used. Spare tires, as well as rocks, logs, or other natural material, are excellent spacers.

Figure 2.14 Neat storage, color coding, and pull cords make these cribbing blocks and wedges easy to use. *Courtesy of Elmer Reese.*

Blocks and Tackle

Because of their mechanical advantage in converting a given amount of pull to a working power greater than the pull, blocks and tackle are useful for lifting or pulling heavy loads. A block is a wooden or metal frame containing one or more pulleys called sheaves (Figure 2.15). Tackle is the assembly of ropes and blocks used to multiply the pulling force.

Simple tackle is one or more blocks reeved with a single rope. Compound tackle is two or more blocks reeved with more than one rope. Because rescuers are most likely to use simple tackle, that is what will be discussed here.

The common terms used in reference to a block and tackle arrangement are standing block, running block, leading block, and fall line (Figure 2.16). The standing block is the block attached to a solid support, and from which the fall line (the line to the power) leads. The running block is the one that is attached to the load to be moved. Leading blocks, usually snatch blocks, change the direction of pull without affecting the mechanical advantage. They are used when the fall line leaves the standing block in such a way that applying the motive force is difficult.

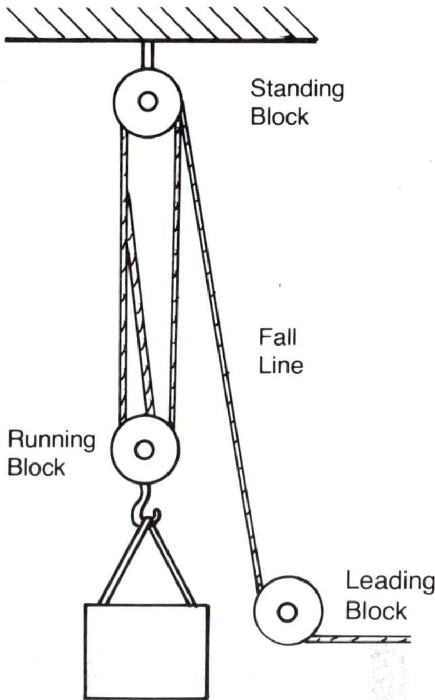

Figure 2.16 Block and tackle arrangement.

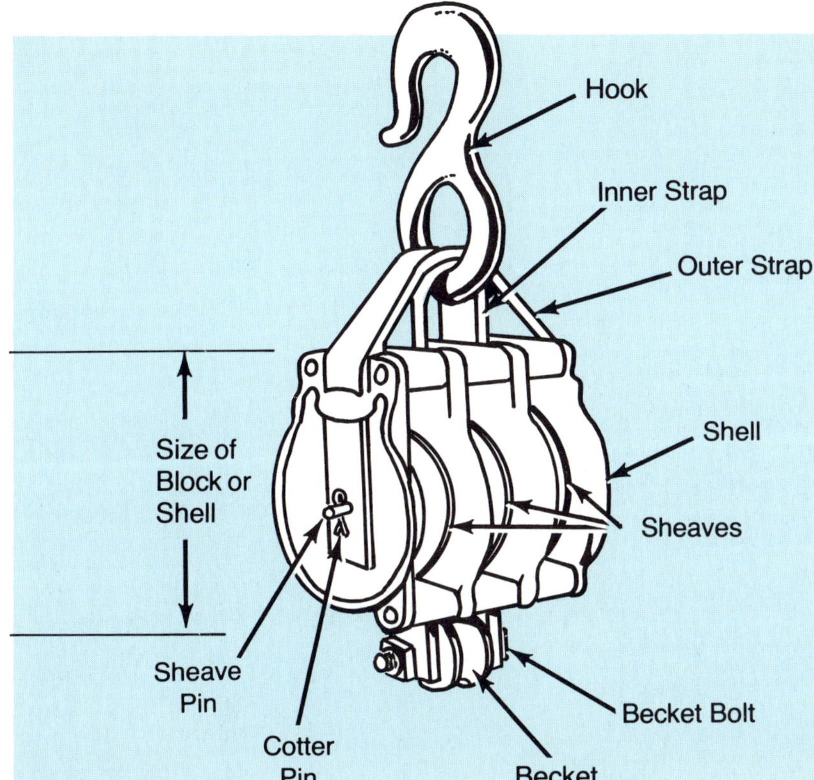

Figure 2.15 Parts of a block.

The methods of reeving single, double, and triple blocks are shown in Figures 2.17 - 2.19. After reeving the blocks, the rope should be pulled back and forth through the blocks several times so the rope will adjust to the blocks. This will help to prevent the rope from twisting under load. Twisting increases friction, chafes the rope, and may jam the blocks. A method of preventing the running block from twisting when traveling horizontal with the ground is shown in Figure 2.20.

PRECAUTIONS FOR USING BLOCKS AND TACKLE

- Be sure rope is the right size for the weight being lifted.
- When pulling on the fall line, everyone should exert a simultaneous steady pull and hold onto the gain.
- Be sure the support holding the standing block will hold the load and the pull.

- Pull in a direct line with the sheaves.
- Whenever possible, the pull should be downhill.
- Pullers should stand so they will be out of danger if the tackle or support fails.
- Easing off on a suspended weight should be done gradually, without jerking.

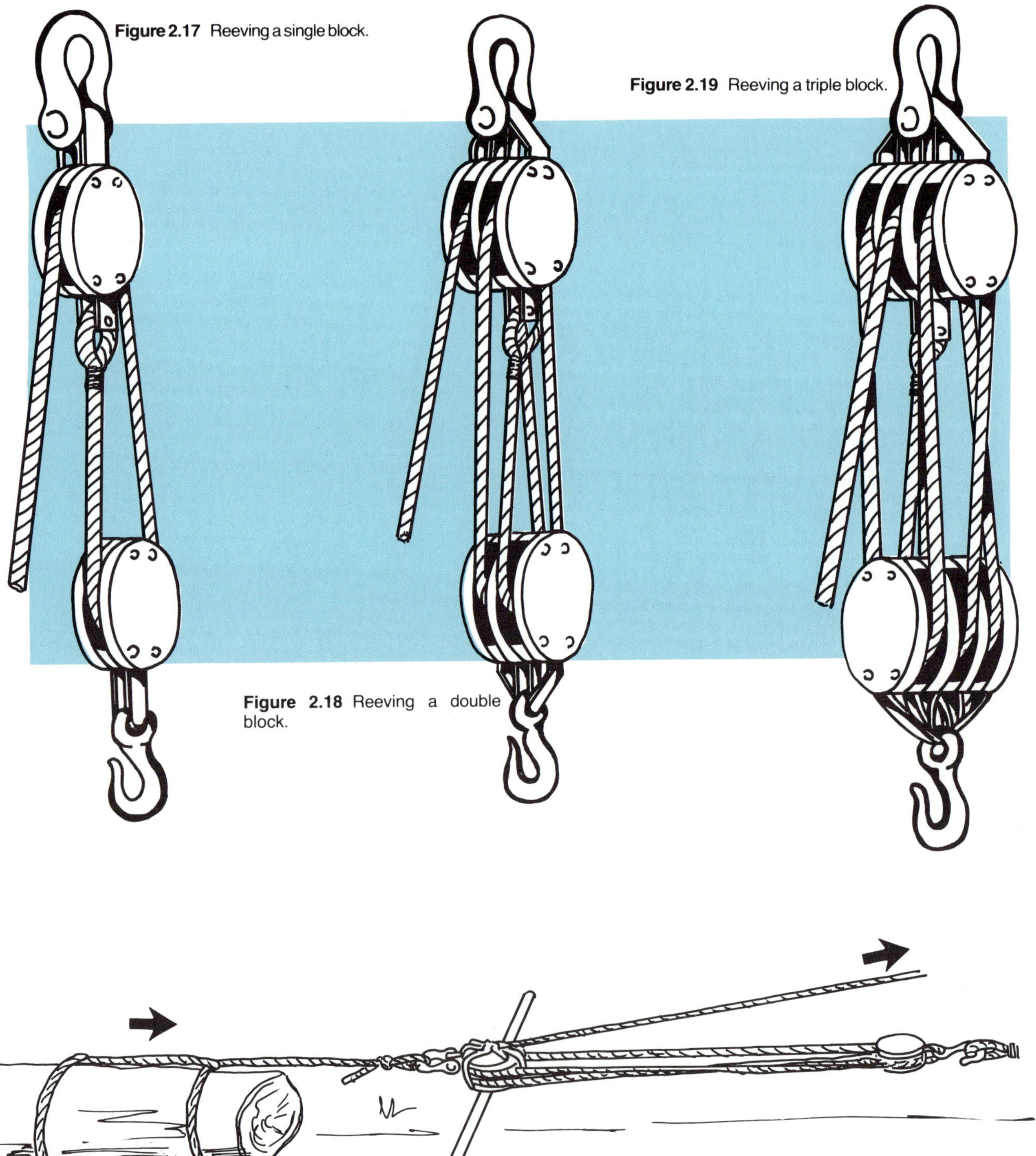

Figure 2.17 Reeving a single block.

Figure 2.18 Reeving a double block.

Figure 2.19 Reeving a triple block.

Figure 2.20 Method of preventing twisting of the running block.

Mousing

Hooks should always be "moused" (Figure 2.21) to prevent slings or ropes from slipping off the hook. Wrap wire or strong twine eight to ten times around the hook, then take several turns around the mousing and tie the ends securely.

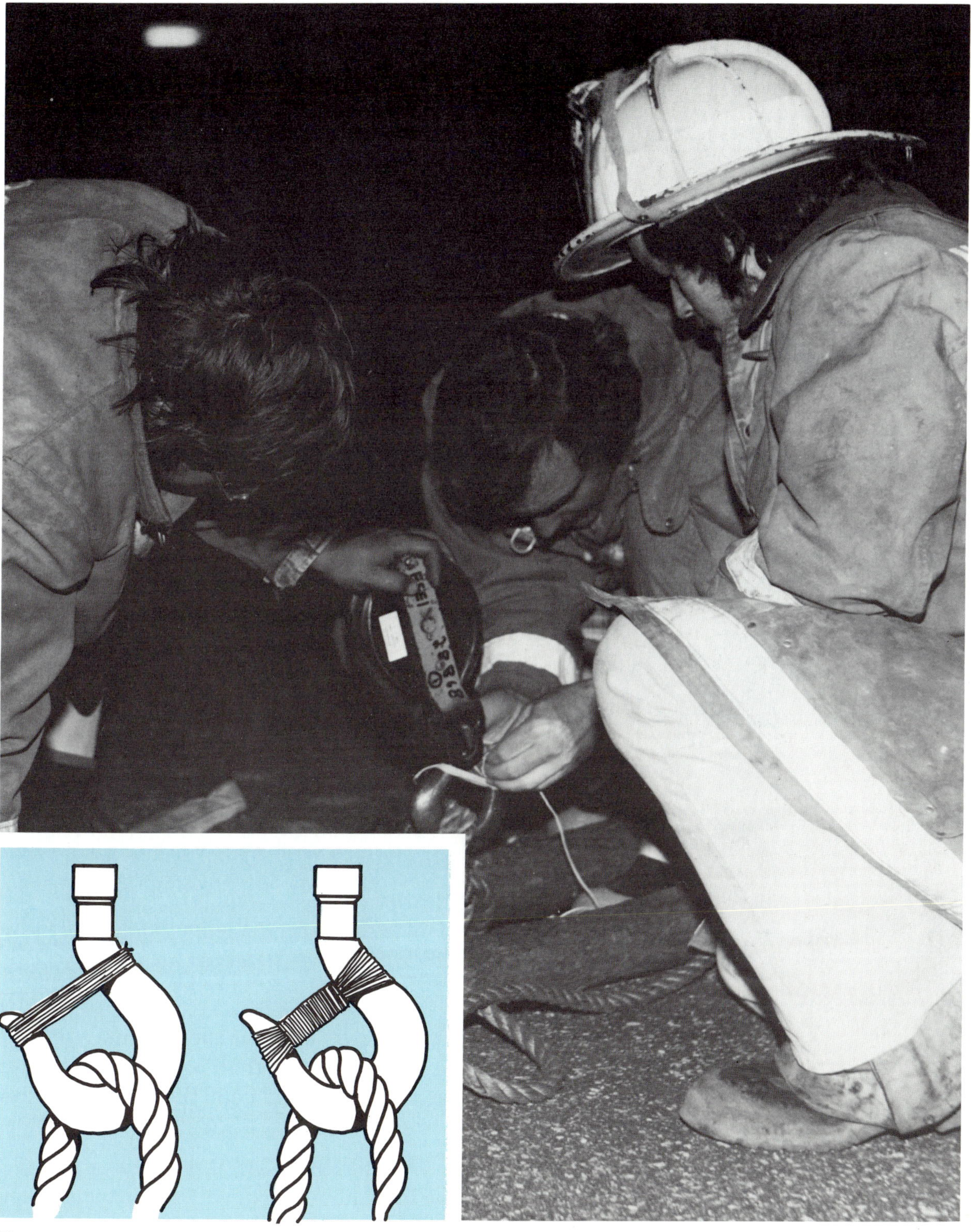

Figure 2.21 Hooks used for hoisting or pulling loads should always be "moused." *Courtesy of Bill Maggi.*

Anchoring Guy Lines

Guy lines should always be fastened to anchors as near to the ground as possible and should leave the anchorage as nearly parallel to the ground as possible to avoid pulling the anchor from the ground (Figure 2.22).

Figure 2.22 A 1-1-1 picket hold-fast being finished. Note angle of ropes. *Courtesy of Bill Maggi.*

Two or more anchors linked together are better than only one anchor. When linking one anchor to another, a point high on one anchor should be linked to a low point near the ground on the anchor behind it.

Stakes used for pickets should be at least three inches (77 mm) in diameter and five feet (1.5 m) long, with three to four feet (1-1.2 m) of the stake driven into the ground. One such anchor is safe for loads of 700 pounds (317 kg) or less.

Multiple-picket holdfasts are made with a number of stakes three inches (77 mm) in diameter and five feet (1.5 m) long. The pickets should be driven about three to six feet (1-2 m) apart. The principal strength of a multiple-picket holdfast is the front picket.

Several kinds of anchors are shown in Figures 2.23 - 2.26.

Figure 2.23 Using two trees for a natural anchor.

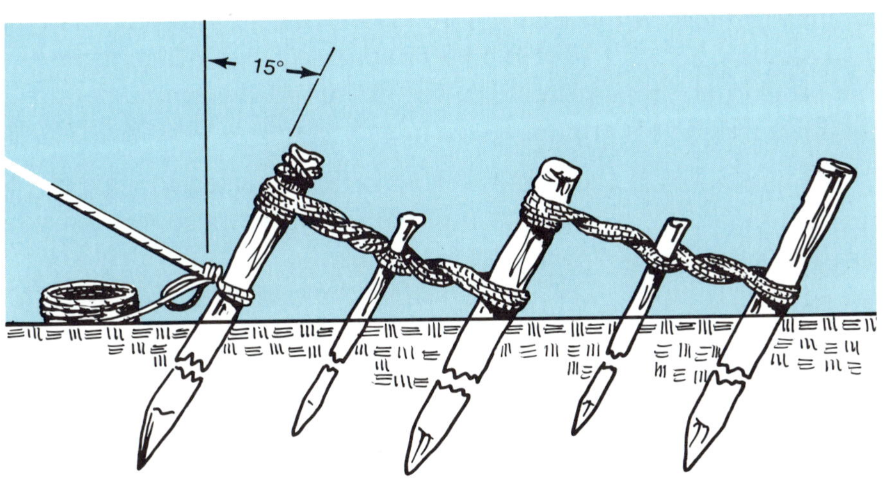

Figure 2.24 A 1-1-1 picket holdfast.

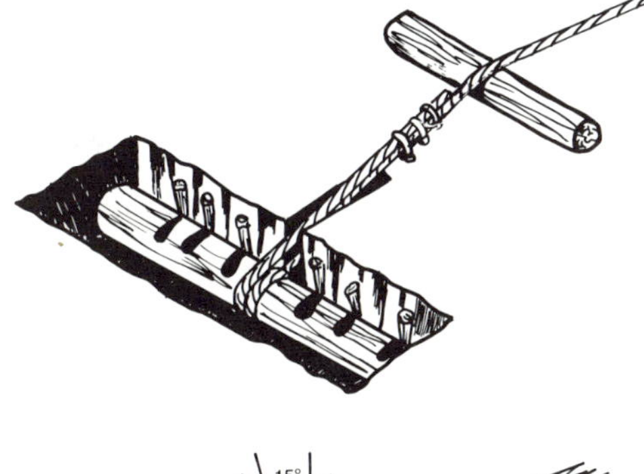

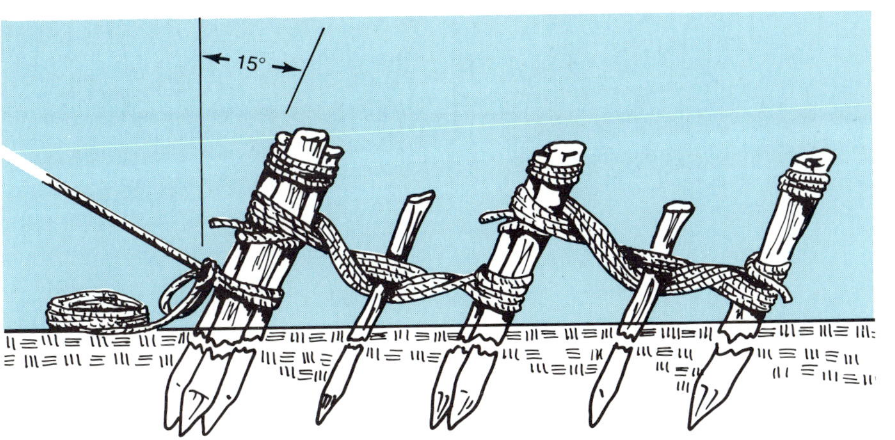

Figure 2.25 A 3-2-1 picket holdfast.

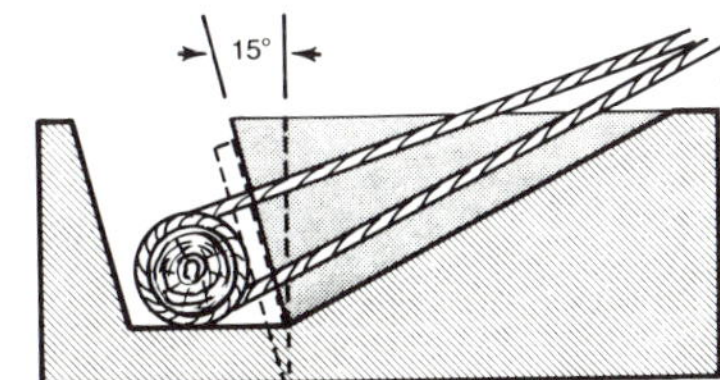

Figure 2.26 A log deadman.

Table 2.1

Holding Power of Picket Holdfasts in Loamy Soil

Holdfast	Pounds	Kilograms
Single picket	700	310
1-1 picket holdfast	1400	630
1-1-1 picket holdfast	1800	810
2-1 picket holdfast	2000	900
3-2-1 picket holdfast	4000	1800

Note: Holding power is much less in wet ground.

Source: U.S. Department of the Army, *Rigging* (Technical Manual TM 5-25) (Washington, D.C.: U.S. Government Printing Office, 1964).

Rigging for Blocks and Tackle

Using blocks and tackle as lifting devices can be aided by rigging such as jibs, derricks, tripods, and A-frames. The straighter a load is picked up, the less energy is required. It may be necessary because of obstacles or space constraints to move a load other than straight up or down. The standard complement of ladders carried will provide materials for rigging.

The gin pole is a single standing pole supported by rope guy lines (Figure 2.27). The pole is angled over the load and secured by the guy lines. The hoisting rope is threaded down the pole to keep the pole in compression and minimize lateral sheer tension. A straight ladder makes an excellent gin pole (Figure 2.28).

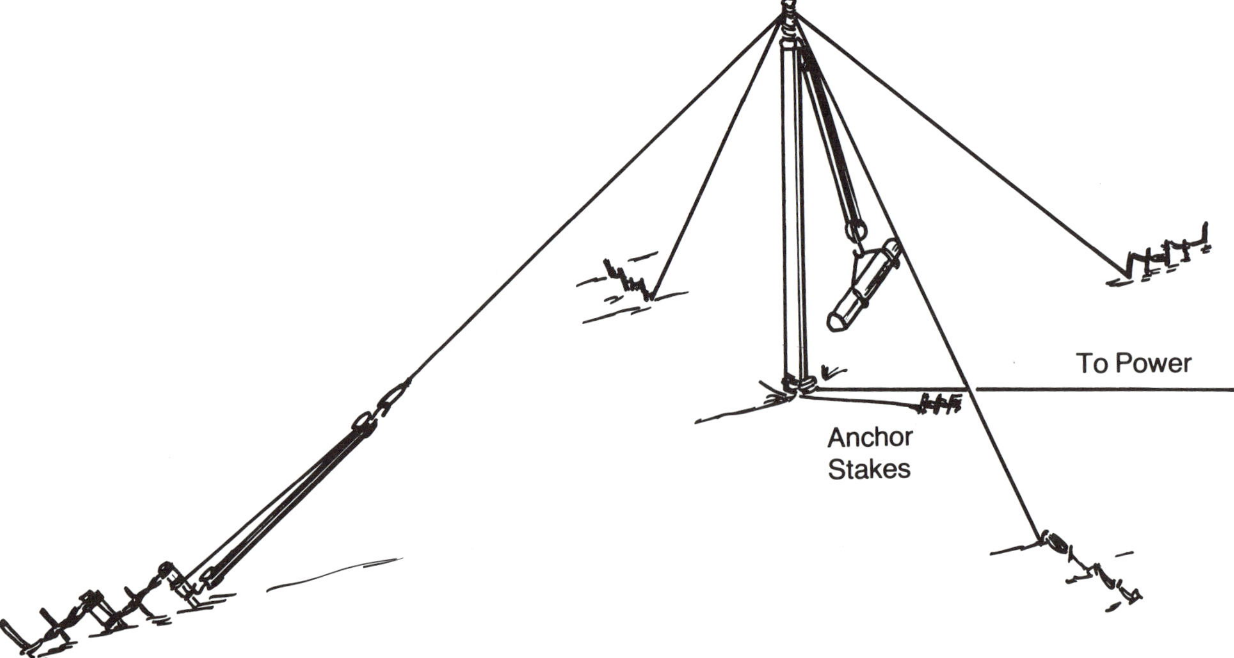

Figure 2.27 A standard gin pole.

Figure 2.28 A gin pole made with a straight ladder.

Erecting a gin pole with a straight ladder requires one rope in addition to the hoisting rope and equipment, several tie-down stakes of suitable length for the load, and a straight ladder. The ladder is laid at the selected site for the gin pole, the butt next to the hoisting site. The additional rope is played out and doubled to form two equal lines with the bend at the tip of the ladder. Approximately three feet (1 m) from the bend, on each of the two lines, a standing bight is taken large enough to slip over the tips of the beams. The loop between the bights is turned to form a figure eight and the slack pulled to the outside loop, which is dropped over the top of the ladder. The line on the opposite side of the ladder is grasped and pulled out, and the figure eight returned. After the procedure has been completed three times, all of the lines on the opposite side of the ladder are pulled forward and the clasp of the standing pulley clipped over them. The hoisting rope is then run to the butt of the ladder and under the bottom rung. The guy line tie-down stakes are driven into the ground a ladder's length from the butt and half a ladder's length to the sides. The ladder is then raised into position and the guys secured.

The ladder gin pole can be improved slightly by adding a second ladder lashed at the top opposite the gin pole. The guy lines are moved to the sides, and the ladders are lashed together across the distance between them at the bottom to prevent them from moving away from each other.

If guy stakes or tie-downs are not available, a third ladder may be placed between the other two with the tip and butt equidistant from the center.

HYDRAULIC TOOLS

Hydraulically operated tools (Figures 2.29-2.34) have had increasing use in rescues. The compression of the hydraulic fluid is done by hand pump, compressed air, or by electricity- or gasoline-powered compressors. Some models can use more than one of these compression methods. The hydraulic rescue tool is versatile, with many having different parts for cutting, spreading, or lifting.

Secure objects before cutting

Because of its great power, the tool, when used for cutting, imparts a large amount of energy to objects it cuts. Therefore, it should not be used to cut anything that is not secured at both ends. Such items include safety door bolts, already cut roof columns, steering wheel segments, and steering columns.

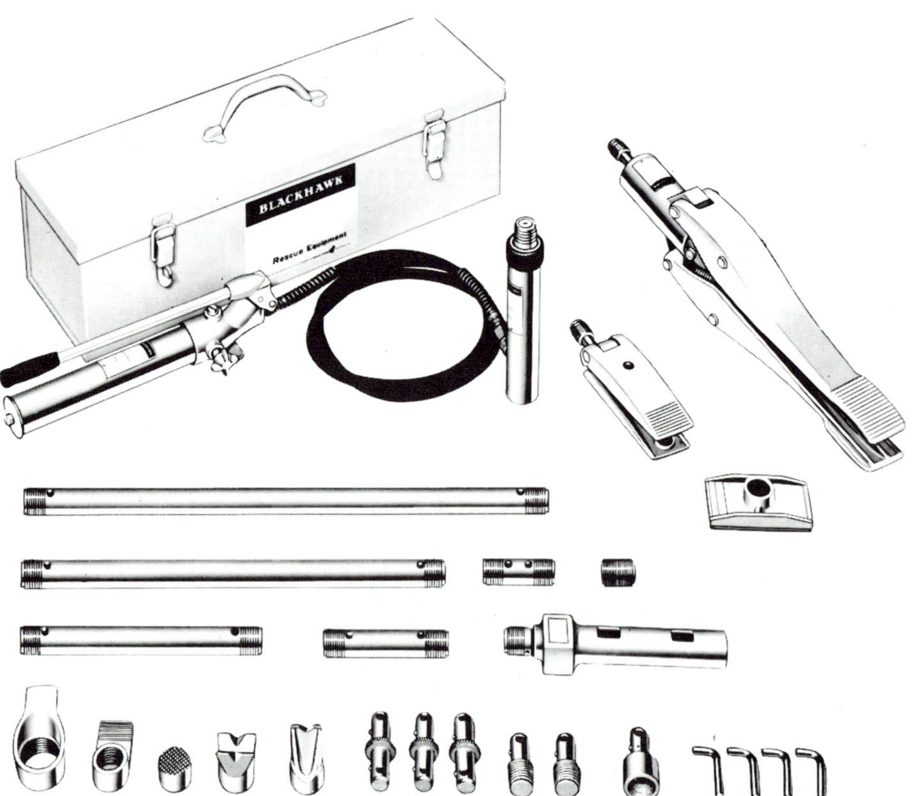

Figure 2.29 One type of hand-powered hydraulic rescue tool whose primary uses are spreading and lifting. Among other jobs it can be used to lift objects, open doors, and brace walls. *Courtesy of Applied Power, Inc.*

Figure 2.30 Hydraulic rescue tools often have various attachments for different jobs, making them versatile. *Courtesy of Hurst Performance.*

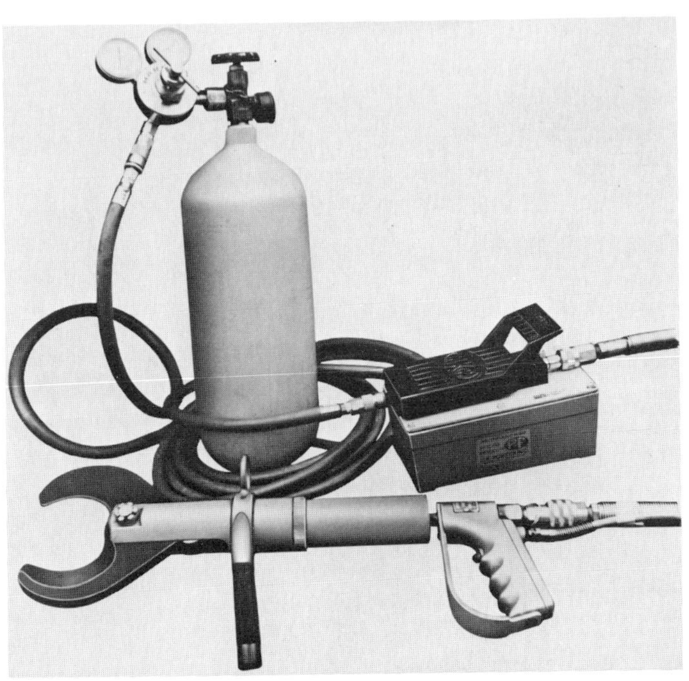

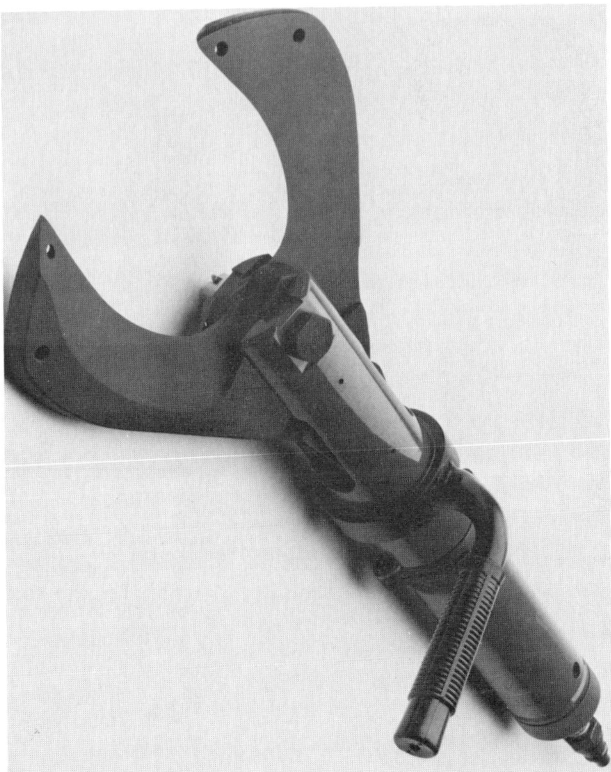

Figure 2.31 A hydraulic rescue tool with a pistol grip and cutting pincers. *Courtesy of H. K. Porter.*

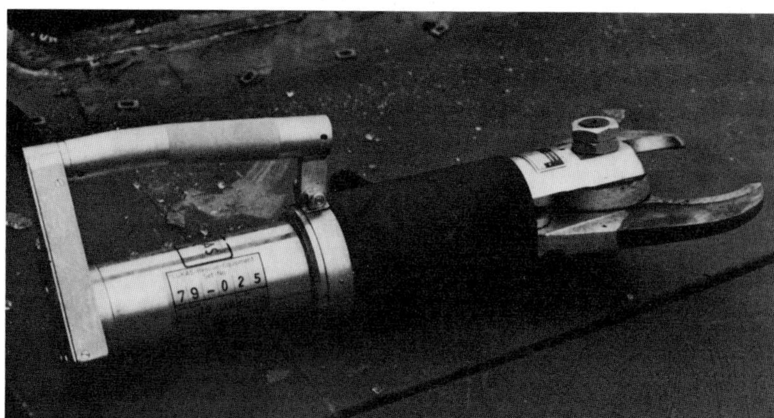

Figure 2.32 The cutting pincers of another model of hydraulic rescue tool.

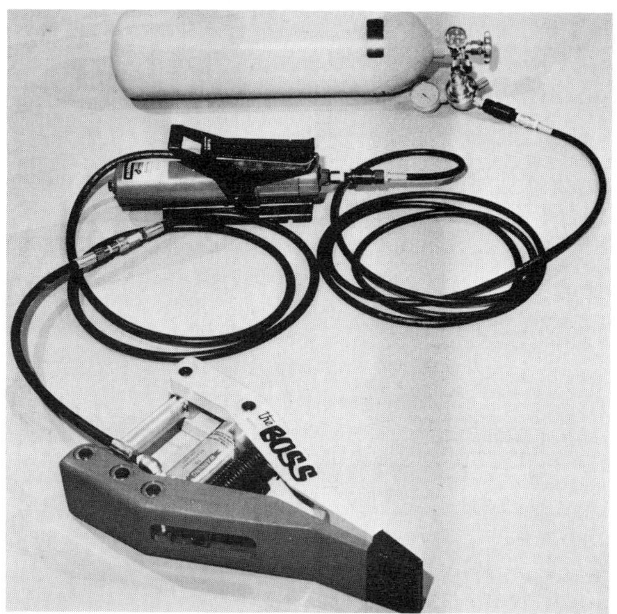

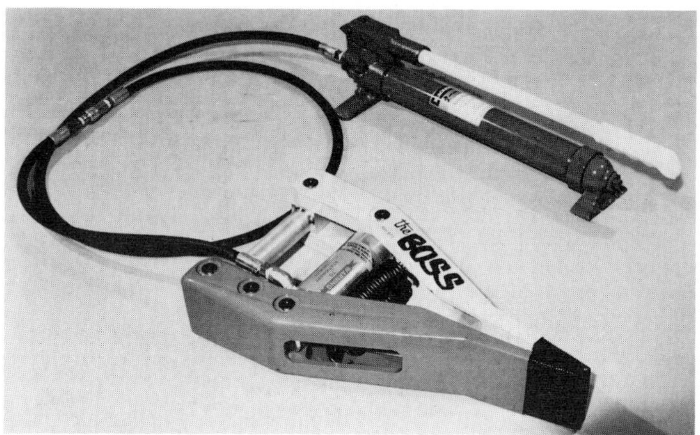

Figure 2.33 This hydraulic rescue tool can be powered by either a hand pump or compressed air. Note the narrow tip that gives good access to small apertures. *Courtesy of Special Service and Supply.*

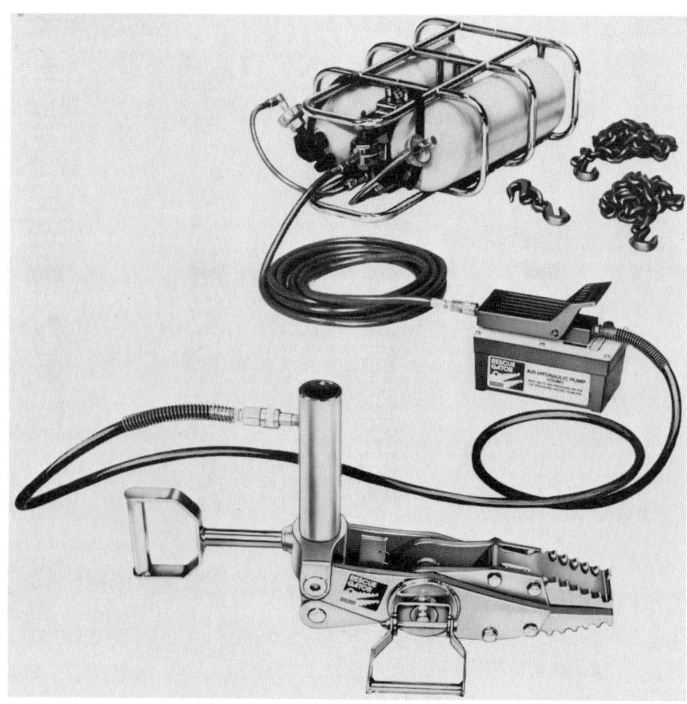

Figure 2.34 Another hydraulic rescue tool that is designed to lift, push, and spread. *Courtesy of Walker Manufacturing.*

MULTIPLE-USE AND SPECIAL TOOLS

Many tools can be used for other than their intended purposes and could be called multiple-use tools, but there are some tools designed for more than one task. The hydraulic tool with different working jaws would fall into this category. There are also hand tools designed specifically so that they can be used for a variety of purposes. Examples of these hand tools are shown in Figures 2.35 and 2.36.

Figure 2.35 A multipurpose hand tool with a sliding-ram handle, designed primarily to cut or pry. *Courtesy of Ziamatic Corp.*

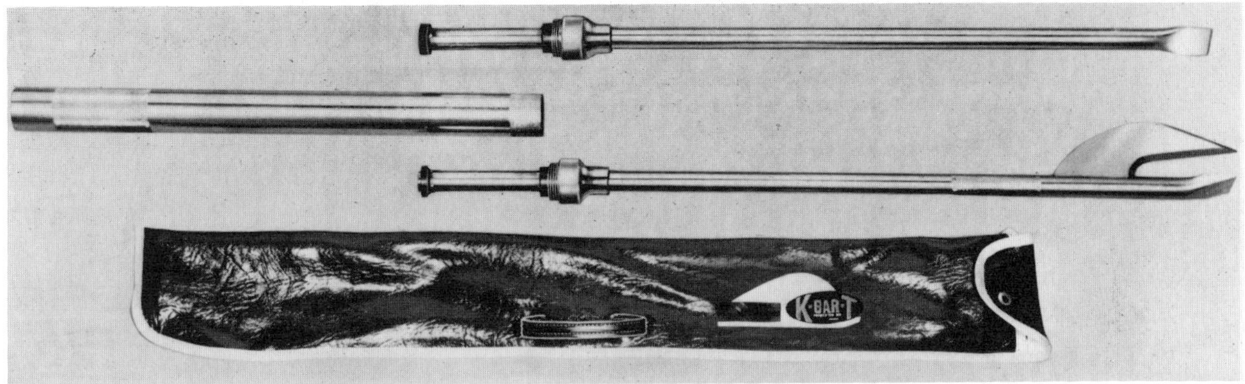

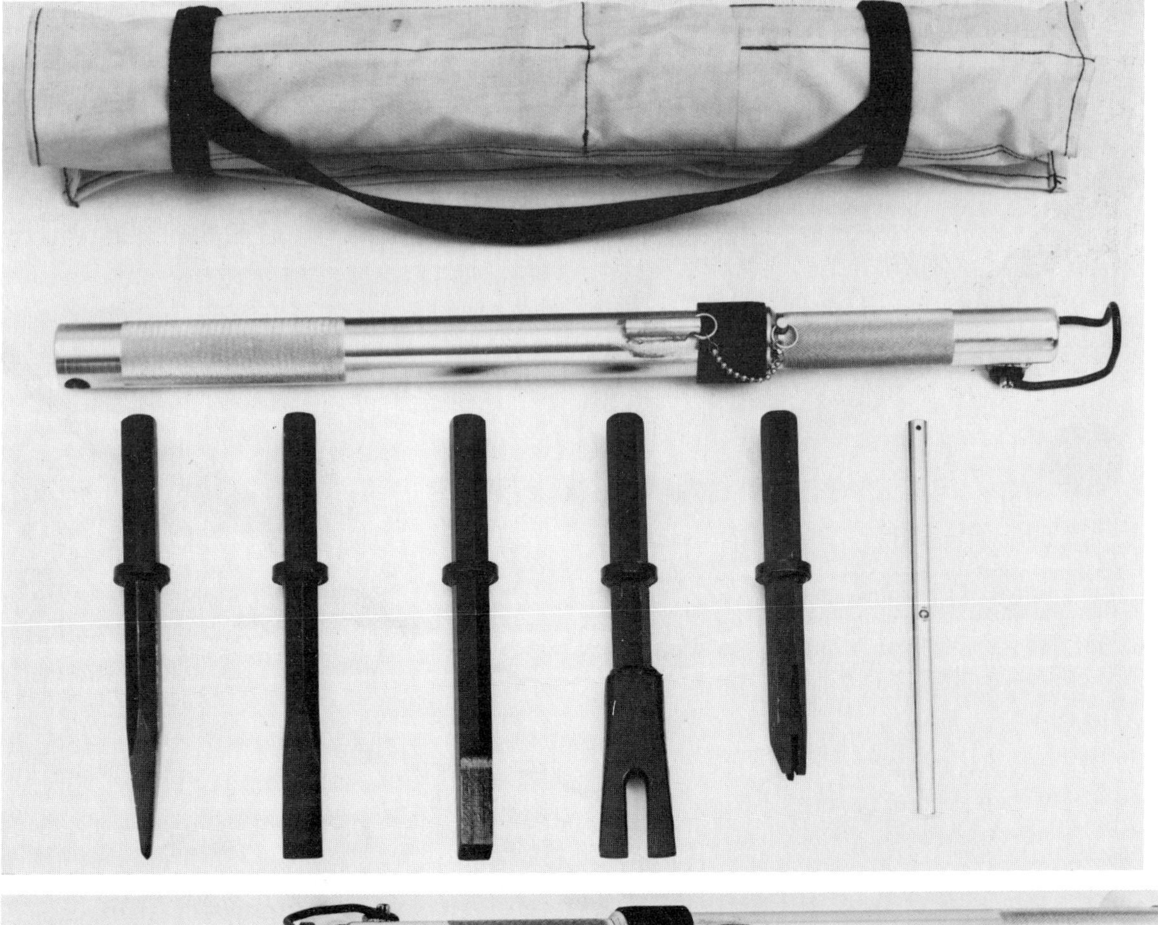

Figure 2.36 This multipurpose striking hand tool has various tool heads and a sliding-ram handle. *Courtesy of Paratech, Inc.*

Homemade tools useful

Rescue personnel with recurring local problems or economic difficulties often have to design their own extrication tools. Some of the homemade tools that have met with some success include cutting tools made with sharpened auto springs and lawn mower blades. They can be used with a hammer or maul like a chisel or axe. They are especially useful in cutting through auto tops and doors. Be sure to pad them to improve the grip.

When chains have to be attached to the underside of the auto, a large hook made of suitable metal will save precious time. Connect the homemade hook to the chain with a shackle, and hook it around the frame. This saves rescue personnel from having to crawl under the car to attach the chain.

Rope And Rescue Knots

Chapter 3

NFPA STANDARD 1001
RESCUE

ROPES/RESCUE
FIRE FIGHTER I

3-5.5 The fire fighter shall demonstrate the methods of inspecting, cleaning and maintaining rope.

ROPES/RESCUE
FIRE FIGHTER II

4-5.1 The fire fighter, when given a simulated fire fighting or rescue task, shall select the appropriate size, strength, and length of rope for the task.
4-5.3 The fire fighter shall use rope to tie ladders, hose, and other equipment so as to secure them to immovable objects.
4-12.6 The fire fighter shall tie a knot and lower a person from a third floor level.

Reprinted by permission from NFPA Standard No. 1001, *Standard for Fire Fighter Professional Qualifications.* Copyright © 1981, National Fire Protection Association, Boston, MA.

Chapter 3
Rope and Rescue Knots

Rope is one of the most versatile rescue tools. It can be used for anything from a lifeline to a means for hoisting, lowering, anchoring, rigging, or even controlling crowds. Pulleys or a more complex block-and-tackle setup greatly increase the rescuer's lifting ability. The right combination of pulleys can multiply the rescuer's lifting ability six times. The rope itself must be of high quality to withstand the stresses such use would cause.

Rope or lines for rescue operations fall into two general categories:

Two types of rescue rope

- Rescue lines (which include lifelines for sliding or rappelling as well as rescue lines for hoisting or lowering) that support human life.
- Utility lines that can be used to transfer small tools to different levels or other light duty in which knots must be tied easily.

The rescue line must, of course, have sufficient strength to support two people or about 600 pounds (272 kg), a 15:1 safety factor or a *minimum* breaking strength of 9,000 pounds (4,081.5 kg) and sufficient diameter for gripping, between five-eights inch (15.9 mm) and one inch (2.54 cm). The utility line, being intended for lighter duty, can be three-eights inch (9.5 mm) to five-eights inch (15.9 mm) in diameter.

For more specific information refer to IFSTA's **Forcible Entry, Rope, and Portable Extinguisher Practices.** Also see Appendix D.

Fire service ropes may be of natural or synthetic fibers. The most common natural fiber is manila, but synthetic fiber ropes are finding increasing application in rescue operations because of their greater strength and resistance to mildew and rot.

Color-coded end whipping can be used to identify kinds and lengths of rope. Whipping is covered in **Forcible Entry, Rope and Portable Extinguisher Practices.**

Rope should be inspected thoroughly after every use. Carefully examine the entire surface for cuts, abrasions, acid damage, or brown spots. Each broken fiber reduces the strength somewhat. Charring or melting from direct flame contact would certainly be damaging, but even exposure to heat will damage a rope, leaving it dry and brittle. Separate the strands at intervals of about three feet (1m). Untwist just enough to expose the interior strands and look for breakage or deterioration of yarns, or the presence of grit. The interior yarns of an overloaded rope will fail first. If the rope has a musty odor, shows brown spots, or separates or breaks easily, the rope should not be used for rescue. Wash only when necessary in cold, clear water, and dry thoroughly to prevent mildew.

ROPE THROW

Many times a rescue rope or lifeline must be thrown from the ground to upper stories of buildings into windows or onto roofs. It might also have to be thrown over downed power lines to remove them from the area. Firefighters have sometimes tied heavy objects such as tools to provide the required weight. The rope throw shown here provides a safe method of getting the rope to upper reaches by allowing only the rope to serve as the weight.

The rope is coiled to the desired weight in either hand (Figure 3.1). The size of these coils will be determined by the size of opening through which the rope will be thrown. Make two wraps around the coils with the standing part (Figure 3.2) and then use

Figure 3.1 The first step in the rope throw is to coil some of the rope to the desired weight in one hand.

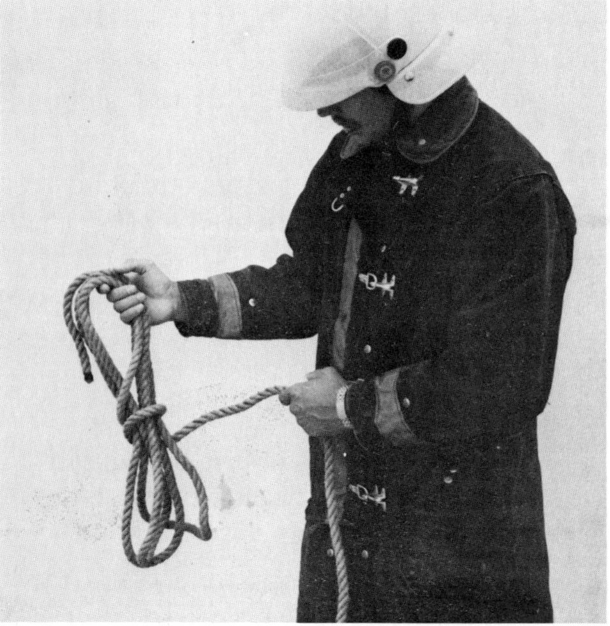

Figure 3.2 Two wraps are made around the coil.

the standing part to form a bight. Tuck the bight through the top of coils (Figure 3.3) to form a handle.

A second series of coils is then formed to provide enough rope for the traveling distance as the rope is thrown to the objective (Figure 3.4).

The rope is then thrown to the objective (Figures 3.5 and 3.6).

Figure 3.3 With the standing part form a bight and pass it through the coil.

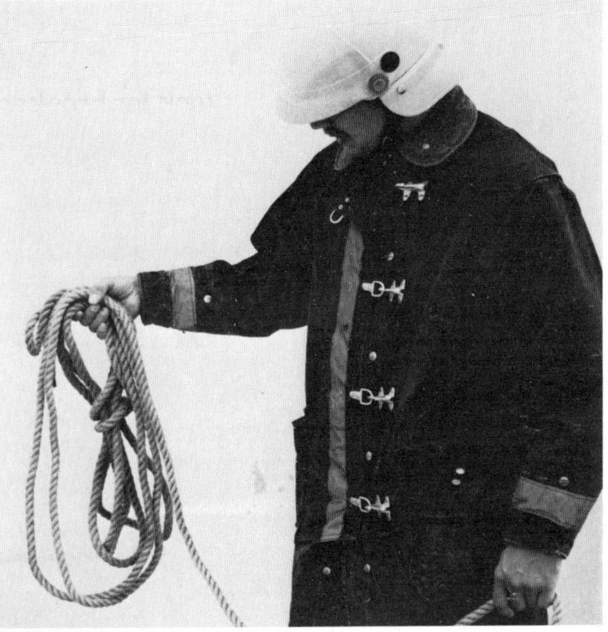

Figure 3.4 The amount of rope needed to reach the objective is made into another coil.

Figures 3.5 and 3.6 The rope is then thrown underhand to the objective.

The coils will fall out upon reaching the goal and the rope is ready for use.

The height and distance achieved in this throw will depend on the physical ability of the thrower and the weight of the rope.

RESCUE KNOTS

Many knots are used in the fire service and all may be used in rescues, but only rescue knots will be discussed and illustrated here. If more information is desired, the reader may refer to two other IFSTA publications: **Essentials of Fire Fighting** and **Forcible Entry, Rope and Portable Extinguisher Practices.** Rescue knots may be used for lowering or raising victims, and other knots may be used to secure objects for raising and lowering during a rescue.

Always use a safety line

CAUTION

A person's injuries usually cannot all be discovered in the field. Therefore, use the following rope techniques for lowering or hoisting only when an injured or unconscious person is in imminent danger and backboards or stretchers are not available. Be sure to use an additional line for a safety line.

Also use a safety line if using firefighters as "victims" in practice.

The Bowline On A Bight

A rescue knot that every firefighter should know for rescue and personal safety is the bowline on a bight. This knot is also commonly known as the life knot. It forms two loops that can be placed around a victim's thighs. An additional loop around the victim's chest secures the life knot, but to avoid pinching or choking the victim it must be tied in a knot.

Step 1: Measure the amount of rope that will be needed. A quick way to make this measurement is done in two steps as shown in Figures 3.7 and 3.8. The length of the rope will then be equal to twice the length of a person's armspread.

Step 2: Double the working end (the end just measured) back upon itself (Figure 3.9).

Step 3: Stand on the doubled rope near the center and, with the doubled rope in each hand, adjust it to the hip or waist height of the rescuer (Figure 3.10). (If the victim or rescuer is exceptionally tall or short, the measurement will have to be adjusted.)

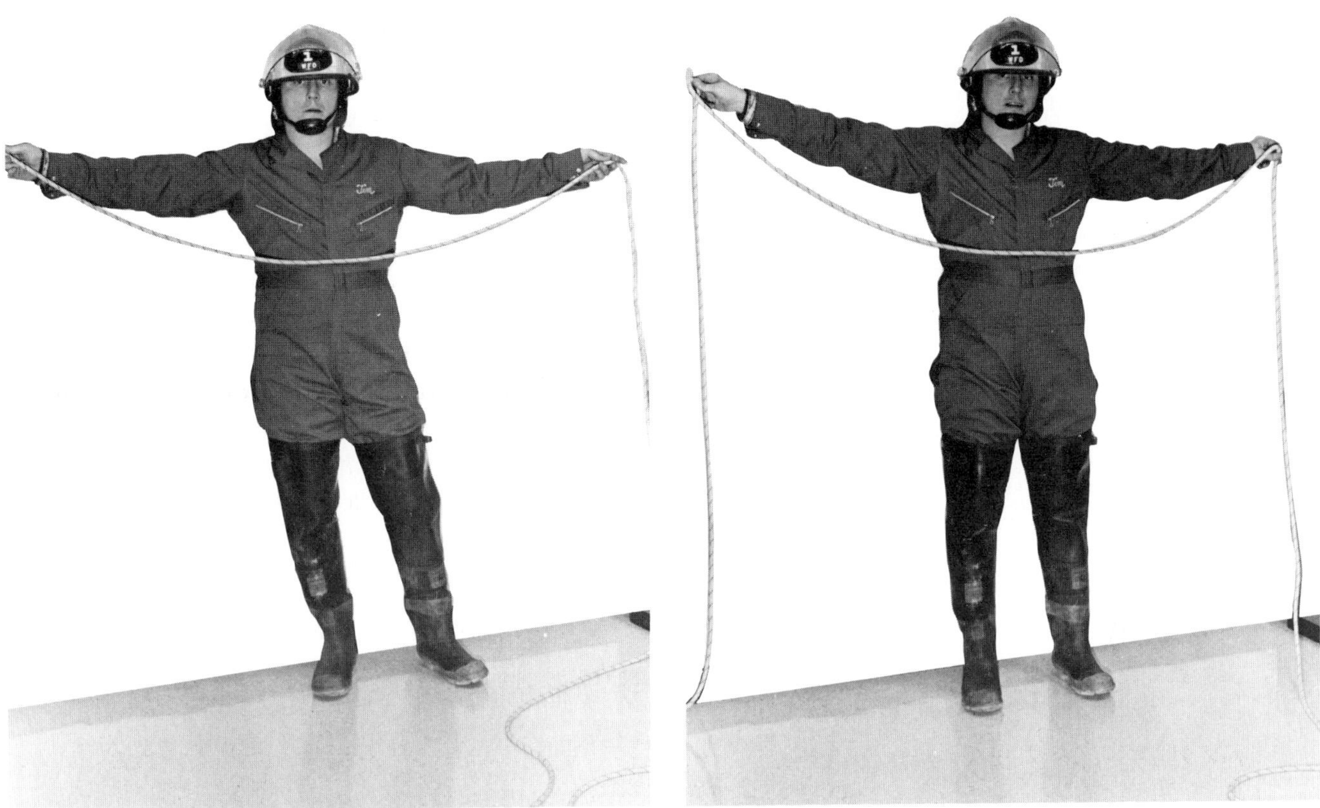

Figures 3.7 - 3.8 Grasp the end of the rope, extend it to full arms spread, then repeat the process so that the measured portion of the rope equals twice the length of your arm spread.

Figure 3.9 Double the measured part of the rope back upon the standing part.

Figure 3.10 Stand on the doubled line near the center and adjust it to the hips or waist.

Step 4: The loop end of the doubled rope is the working end. Tie an overhand knot to form a loop about three feet (1 m) long, so the loop hangs downward through the overhand knot (Figure 3.11).

Step 5: Hold the overhand knot in one hand, reach down with the other hand, and bring the three-foot (1 m) loop up and over the hand holding the knot (Figure 3.12).

Figure 3.11 Tie a loose overhand knot to form a loop about three feet long.

Figure 3.12 Bring the loop up and over the hand.

Step 6: Place the loop in one hand under the thumb of the other. Grasp the two sides of the loop where they go through the overhand knot and pull upward to form a bight (Figure 3.13).

Step 7: Tighten the bight and adjust the two loops so they are even (Figures 3.14 and 3.15).

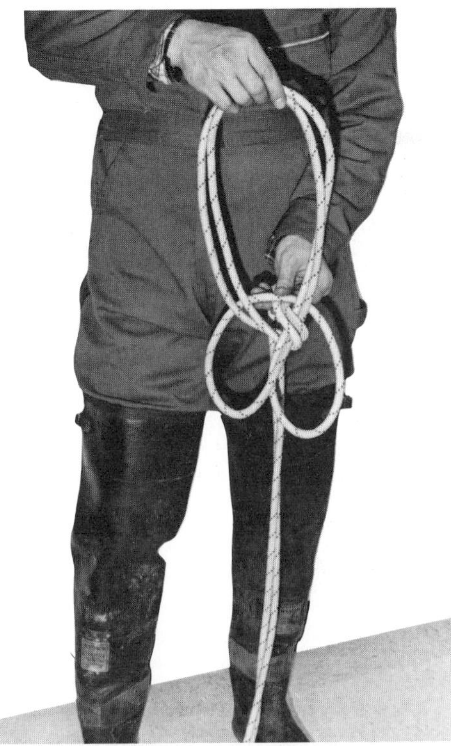

Figure 3.13 Grasp the two sides of the loop where they go through the overhand knot and pull upward to form a bight.

Figures 3.14-3.15 Tighten the bight and adjust the two loops so they are even.

Step 8: Put the victim's legs through the two loops and draw them well up in the crotch (Figure 3.16).

Step 9: Place a half hitch around the victim's chest with the working line and secure it in front with an overhand safety to form a loop (Figure 3.17).

Step 10: Place the loose end of the line through the loop that was just formed, and pull on the working line to secure the tie (Figure 3.18).

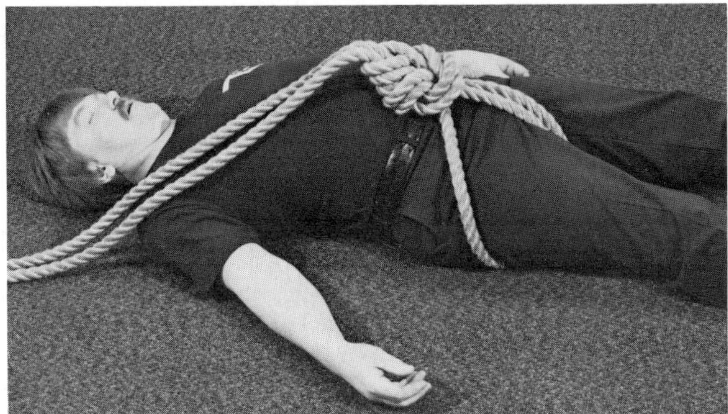

Figure 3.16 Put the victim's legs through the loops and draw the loops well up in the crotch.

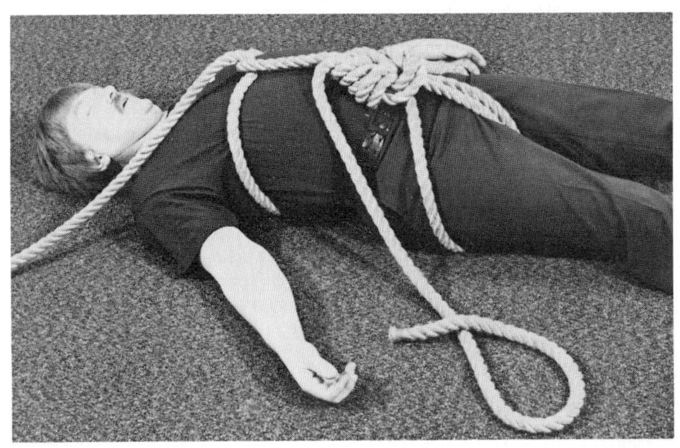

Figure 3.17 Place a half hitch around the victim's chest with the working line and secure it in front with an overhand safety to form a loop.

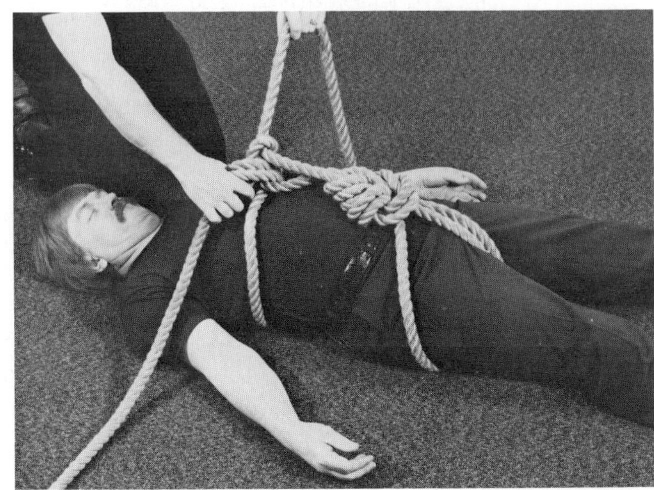

Figure 3.18 Place the loose end of the line through the loop, and pull on the working line to secure the tie.

The Three-Loop Bowline

This bowline makes three loops in a rope to form a sling that can be easily placed on an unconscious or injured victim. One loop fits around each leg and the third loop goes around the chest. The rope is doubled over on itself, forming a long bight used as the running end of the rope to tie a bowline knot.

Step 1: Double the working end of a line back upon itself about eight feet (2.4 m) as for the bowline on a bight. Tie a single bowline in the double line about two-thirds from the working end. Before the knot is tightened, adjust two of the three loops so they will be about six inches (150 mm) longer than the other one (Figure 3.19).

Figure 3.19 About two-thirds from the working end of the doubled rope tie a single bowline. Before the knot is tightened adjust two of the three loops to about six inches (150 mm) longer than the other one.

Step 2: Support the victim's feet and legs upon one of the rescuer's knees. Place the short loop over both feet and let it fall about the victim's legs.

Step 3: Grasp the other two loops, one in each hand, and place one loop over each of the victim's feet.

Step 4: With one arm under the victim's legs, rock the lower extremities upward so the short loop can pass toward the armpits.

Step 5: Lower the victim's feet and slide the two long loops to the groin, adjusting them snugly without cutting off circulation.

Step 6: Raise the victim by the shoulders to a sitting position, bring the short loop up to the armpits and across the back.

Step 7: Place the bowline knot next to the victim's sternum and remove the slack from the loops. Hold the knot next to the chest and pull the slack from one side of the chest and one leg. Then repeat the process on the other side (Figure 3.20).

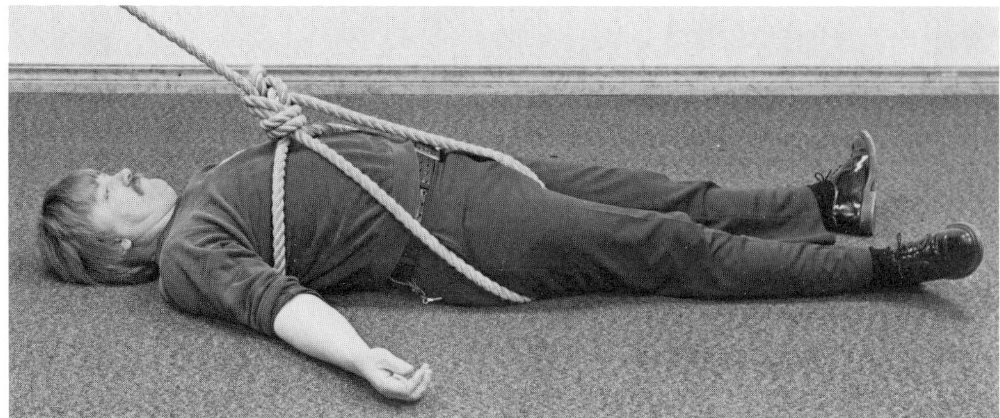

Figure 3.20 The three-loop bowline secured to the victim.

RESCUE SLING

This sling is a versatile device made from a 15-foot (4.5 m) length of ¾-inch (20 mm) rope. It can be used to raise, lower, or carry a victim. The sling is relatively simple to make because it only requires one knot, the bowline.

To make the sling, a bowline is tied at each end of the rope to form loops large enough to fit around the legs (Figure 3.21). The first loop is slid up to the top of one leg and the other end passed under the small of the back. The second loop is slid up to the top of the other leg. The knots are positioned on top of the thighs. The middle section of the rope is then worked up the back to a position under the arms.

Figure 3.21 A bowline is made at each end of a 15-foot (4.5 m) rope. *Courtesy of Salem, Va., Fire Dept.*

To raise or lower a victim, bring together the sides of the loop that passes under the back. Using the end of a hoisting rope, tie them together with a bowline as high up on the victim's body as possible. The victim can then be raised or lowered (Figures 3.22 and 3.23).

To carry the victim, the rescuer uses the small loops formed in the rope, between the knots and the point where the rope passes under the victim, for arm straps. The rescuer positions the victim on one side and lies down with the victim at the rescuer's back. The rescuer's arms are then slid through the straps. The rescuer rolls onto the stomach and the victim will roll on top. The rescuer can then stand up or crawl, carrying the victim.

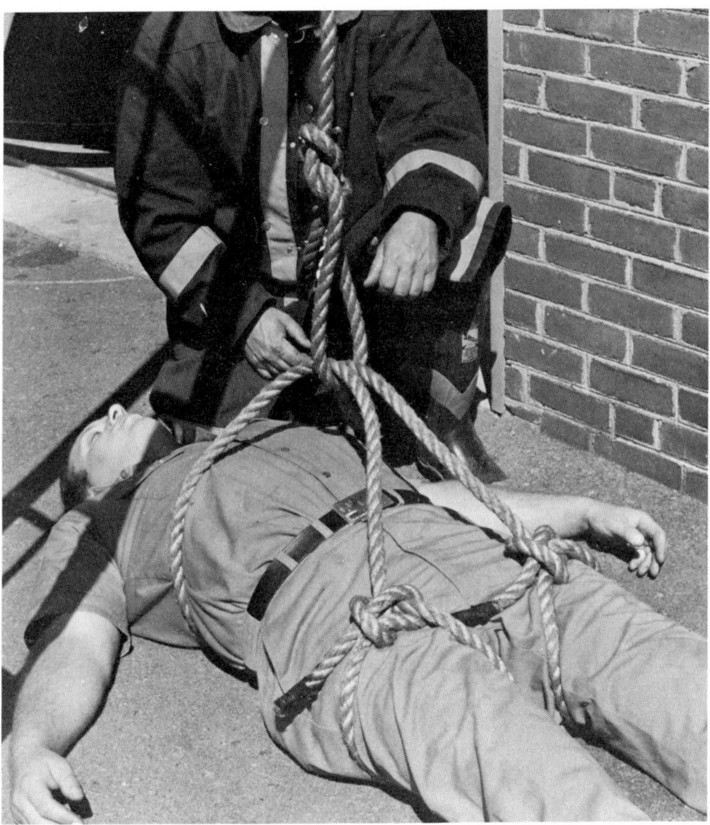

Figure 3.22 The legs are passed through the loops and the rope worked up the trunk of the victim. *Courtesy of Salem, Va., Fire Dept.*

Figure 3.23 The victim can be raised or lowered with this sling. *Courtesy of Salem, Va., Fire Dept.*

Rescue Situations

Chapter 4

NFPA STANDARD 1001
RESCUE
FIRE FIGHTER I

3-15.1 The fire fighter shall identify dangerous building conditions created by fire.

RESCUE
FIRE FIGHTER II

4-12.1 The fire fighter shall demonstrate the procedure to remove debris, rubble, and other materials found at a cave-in.

4-12.2 The fire fighter shall demonstrate the use of the following rescue tools:

(a) Shoring blocks
(b) Trench jacks
(c) Block and tackle
(d) Hydraulic jacks
(e) Screw jacks

4-12.4 The fire fighter shall identify dangers of search and rescue missions in tunnels, caves, construction sites, and other hazardous areas.

Reprinted by permission from NFPA Standard No. 1001, *Standard for Fire Fighter Professional Qualifications.* Copyright © 1981, National Fire Protection Association, Boston, MA.

Chapter 4

Rescue Situations

To determine rescue procedures, a review of the many situations that may exist is necessary. Since incidents cause hazardous situations, rescue personnel may expect to find people involved with any one of many situations. By knowing the situation at hand and being able to anticipate existing conditions, rescue workers are better able to carefully remove victims to a safe area.

Three types of situations

The many possible types of rescue situations are too numerous to list. But the conditions under which those rescue situations will be encountered can be divided into three groups:

Group I: Rescues in which the victim can be reached directly by the rescuer and directed or assisted to safety without undue risk of injury to the rescuer or victim.

Group II: Rescues in which the rescuer knows the victim's location, and the victim is trapped so removal necessarily involves risk of injury to the rescuer or victim.

Group III: Rescues in which the rescuer does not know the victim's location, and the victim is trapped so removal necessarily involves risk of injury to the rescuer or victim.

Each group obviously requires a slightly different series of actions on the part of the firefighter. The exact actions and their sequence will be determined by many factors and considerations, some of which are:

- The amount of danger to the victim and rescuer
- The alternative methods of rescue available
- The personnel available for rescue operations

- The equipment available for rescue operations
- The amount of time available
- The knowledge and experience of the rescuers

RESCUE FROM BURNING BUILDINGS

Correct suppression techniques important

The fire department arriving at a structure fire faces two primary objectives: saving lives and saving property, generally in that order. However, both objectives often must be addressed simultaneously. The use of correct suppression and ventilation tactics will go a long way in effecting a safe rescue. Proper ventilation will eliminate the hazards of backdraft, remove hot toxic fire gases, and increase visibility by reducing dense smoke. Proper hose placement will separate the victim from fire spread, insure avenues of escape, and provide protection for rescue workers.

Rescuing persons from burning buildings is often difficult. The rescuer must overcome many obstacles, often under the most adverse conditions. The rescuer faces many potentially lethal hazards — fire, toxic gases, dense smoke, and possible building collapse. In addition, the rescuer must function with severe handicaps, such as limited knowledge of the building, insufficient knowledge of victim location, and possible irrational actions on the part of the victim.

The physical problems facing rescuers add a certain degree of difficulty. Rescuers will be working in temperatures well above those normally encountered. They will be forced to carry an appreciable amount of extra weight in the form of protective equipment and breathing apparatus, and may be required to drag or carry an unconscious victim to safety. Under the stress of the situation, the firefighters' own physiological reactions may additionally hamper their physical rescue ability.

Planning important

Effective rescue begins well before the alarm is received. Realistically, pre-fire planning is done only on the larger buildings in a community. However, there are many things the firefighter may do to increase rescue effectiveness. Studying basic building construction provides insight into potential fire and rescue problems. Typical residential room plans can be reviewed and model homes in tract developments can be toured. Also, firefighters should tour the larger buildings in their district when possible.

When the fire department arrives on the scene, pre-fire planning will be a primary tool, but additional help from building occupants will often be necessary. They should be found and questioned about people still in the building. But the prudent fire officer remembers that witnesses at fire scenes are often unreliable sources of information. There are many reported instances of witnesses stating that victims were trapped in a building when in fact there was no one trapped, or witnesses stating that no one was in a building which was in fact occupied. Generally, the only acceptable method of determining if there are persons still in a building is by search. A search of a building should be initiated by the fire department at the earliest possible moment.

RESCUE FROM DEMOLISHED BUILDINGS

The difficulty encountered in reaching a victim in a demolished building depends upon existing conditions. Victims on the surface or lightly trapped should be rescued first. This immediate rescue will take care of injured persons who are not trapped and those persons who are trapped to the extent that their rescue can be accomplished quickly. Rescue of a heavily trapped victim is a more complicated endeavor and requires more time. This type of rescue depends upon the services of trained rescue workers who can use rescue tools and equipment well.

When floor supports fail in any type of building, the floors and roof may drop in large sections and form voids. If these sections remain in one piece, supported on one side but collapsed on the other, they form a lean-to collapse as illustrated in Figure 4.1. Weakening or destruction of bearing walls may cause the floors or the roof to collapse. This collapse may cause the debris to fall as far as the lower floor or basement. A collapse of this nature is called a pancake collapse, illustrated in Figure 4.2. Voids may be

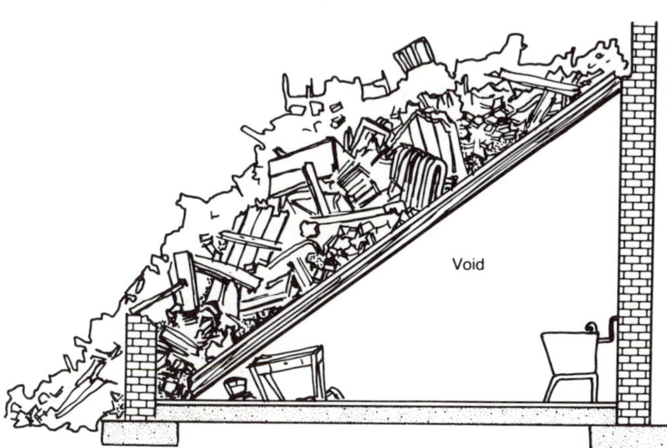

Figure 4.1 Lean-to floor collapse.

Figure 4.2 Pancake floor collapse.

formed between the floors. When heavy loads such as furniture and equipment are concentrated near the center of a floor, the excessive weight may cause the floor to give way. This form is a V collapse, illustrated in Figure 4.3.

Figure 4.3 V floor collapse.

DEBRIS TUNNELING AND SHAFT CONSTRUCTION

Debris tunneling and shaft construction are methods of reaching trapped victims at cave-ins or collapsed buildings, usually when their location is known. Both methods are dangerous and tedious. Progress is often at a rate of only 2½ feet (75 cm) per hour, so tunneling and shaft construction should be considered only after all other methods have been explored. Debris tunneling should be used primarily to connect voids left after collapse. Shaft digging is especially useful when large masses of debris are encountered and persons are trapped in basements or below ground level.

In debris tunneling, firefighters form a crawlway that is safe for rescue workers and large enough to extricate a trapped victim. Tunneling should not be used for general search and must not be aimless. However, in some cases tunneling may be used to reach an area such as a void under a floor so further searching may be done.

Particular attention should be paid to tunnel construction. It should be done from the lowest level possible and without sharp turns. Whenever possible, drive tunnels along a wall or between a wall and a concrete floor to simplify the framing required. Remember that even though the ground may appear solid, the sides of a tunnel must always be supported by timbering. Tunnels 30 inches wide by 36 inches high (76 by 91 cm) have proved satisfactory for rescue work.

Debris tunneling is quite different from tunneling through undisturbed earth. Although strutting and bracing are necessary in both cases, debris tunneling may be irregularly shaped to avoid or properly reinforce unstable or key beams. The kind of material needed to strut and brace a debris tunnel depends on the nature of the job. It can normally be found at the collapse site. Timber from the damaged building may be cut to the lengths needed. Simple props with headpieces are most suitable for supporting large pieces of debris such as collapsed floors, roof timbers, and heavy pieces of masonry. The props must be wedged into position and fixed so they will not slip or become loose and fall out if the debris shifts. But in no case should they be used to jack up the debris. Another method of supporting tunnels is to use frames prefabricated outside the tunnel and installed as the work progresses. Figure 4.4 shows the construction of a frame tunnel. The sides of the tunnel may be lined with boards driven between frame struts.

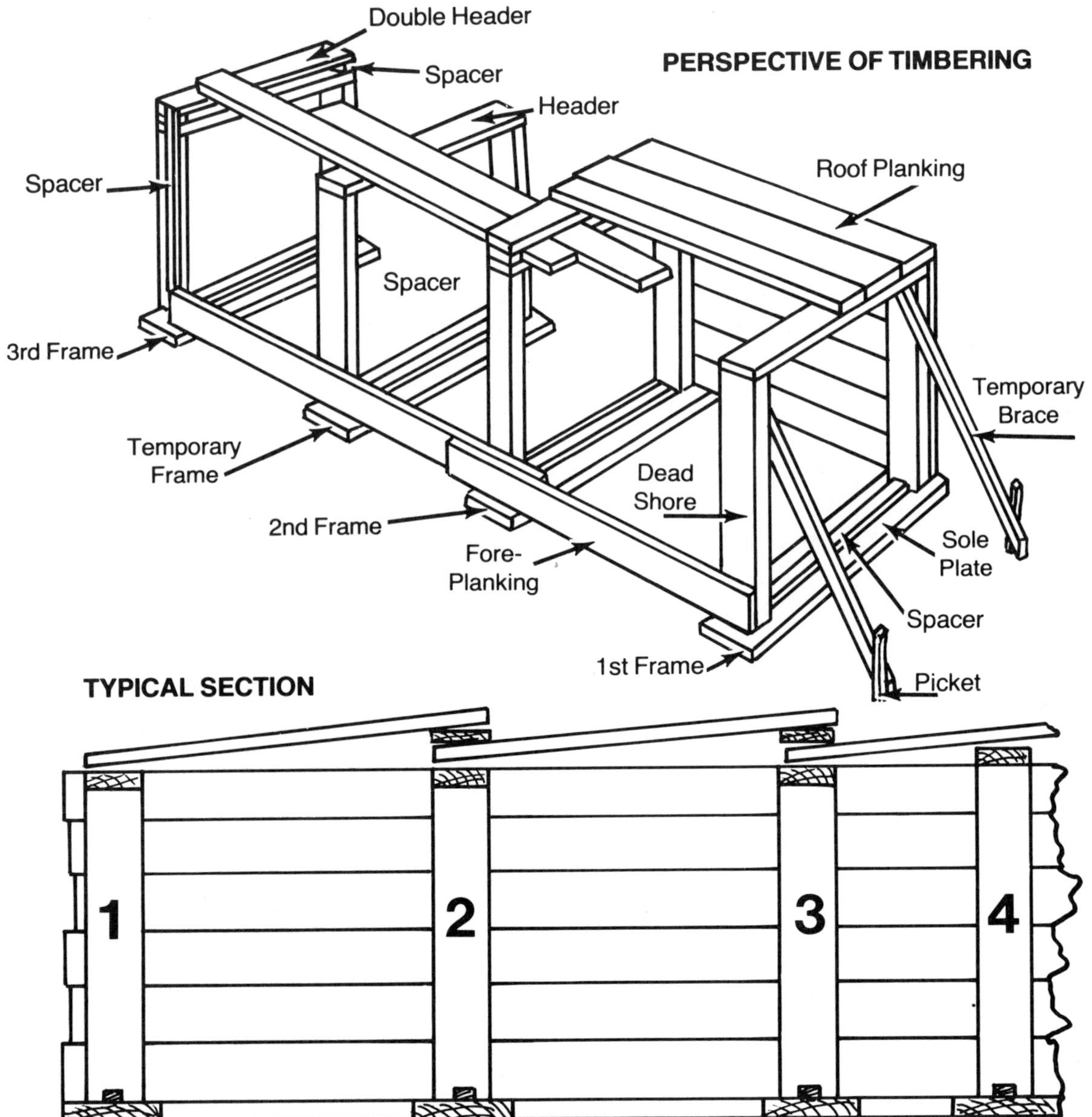

Figure 4.4 Example of tunnel construction for tunneling through debris.

In some circumstances, especially where large amounts of building debris are encountered, the best way to reach persons trapped in a basement is to sink a shaft or hole to the basement level in the ground near the building, and then construct a horizontal tunnel from the bottom of this shaft into the basement (Figure 4.5). This method will eliminate some of the danger and

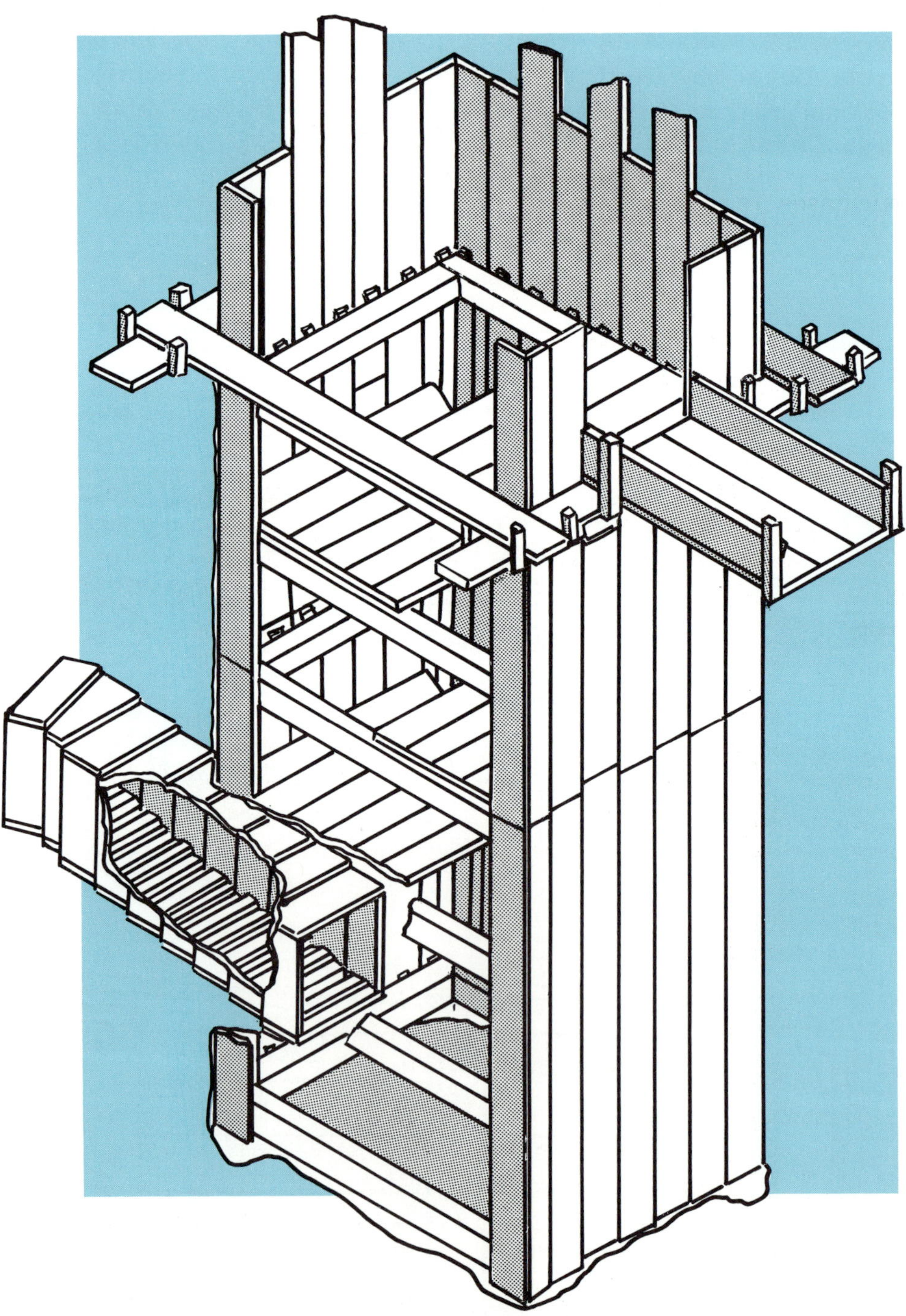

Figure 4.5 Construction of a shaft with a tunnel at the bottom. *Courtesy of Bill Maggi.*

difficulty of debris tunneling. As much of the debris as possible should be removed first so the shaft can be started close to the building. Again, remember that even though the ground may appear solid, the sides of the shaft must always be supported by shoring. Rough and improvised materials for shoring can generally be found on the site. The timbers should be strong enough to support the shaft walls and should be wedged into position as tightly as possible. A shaft should be driven into the earth where there are no service pipes or other obstructions.

Shoring always necessary

During tunneling and shaft construction, personnel should be aware of special hazards that may exist. First, atmospheres in confined spaces may be oxygen deficient or contaminated with toxic gases. In this case, self-contained breathing apparatus is a must. Second, debris is often unstable, and personnel must not be allowed to climb on top of debris during clearing operations unless absolutely necessary. Finally, gas pipes, water lines, telephone cable, and buried electrical conduit may be encountered unexpectedly. Personnel should avoid cutting these lines. If cutting must be done and workers from utility companies are available, let them work on the lines. All electrical lines should be regarded as live unless proved otherwise. The above are certainly not the only hazards. Every rescue operation will have a different set of problems. Rescue personnel must analyze each rescue site for the type of hazards they may expect to encounter.

Tunneling and shaft work should not present any particular difficulties to individuals experienced in excavation. However, when professional help is not immediately on the scene, as will probably be the case, rescue personnel should consider contacting local agencies such as the Army's Corps of Engineers, local civil preparedness, city engineers, or construction companies. These should be able to provide information and assistance.

Know possible sources of aid

TRENCHING

An open trench can often be completed more quickly than a tunnel if debris is not piled too high. Trenching and tunneling operations may sometimes be combined, with a trench extending into the debris until a tunnel becomes more practical.

To trench through debris, start by removing the larger pieces of timber, stones, or other objects from the face of the pile nearest the objective. Then clear a way into the debris by shoveling and other hand methods, removing just enough material to provide a safe passageway.

Trenching may be dangerous. If a trench collapses, the worker has little chance to avoid injury. Use bracing or some other method to retain the sides to prevent the trench from collapsing or the sides from shifting dangerously.

One satisfactory method is to drive sufficient sheet piling (usually lumber found at the site) into the ground. Additional support may be provided by horizontally bracing the sheet piling by putting screw or building jacks, if available, or wood struts between the two retaining walls (Figures 4.6 - 4.20).

Material removed from a trench should be piled some distance from the edge so it will not fall back into the trench or have to be moved again. The size of the trench will be governed by its purpose and the nature of the debris.

Trenching is used to reach a specific point, not for general clearance. Leaders may decide to start two or more trenches to a given point simultaneously, since it is not always possible to determine the fastest route.

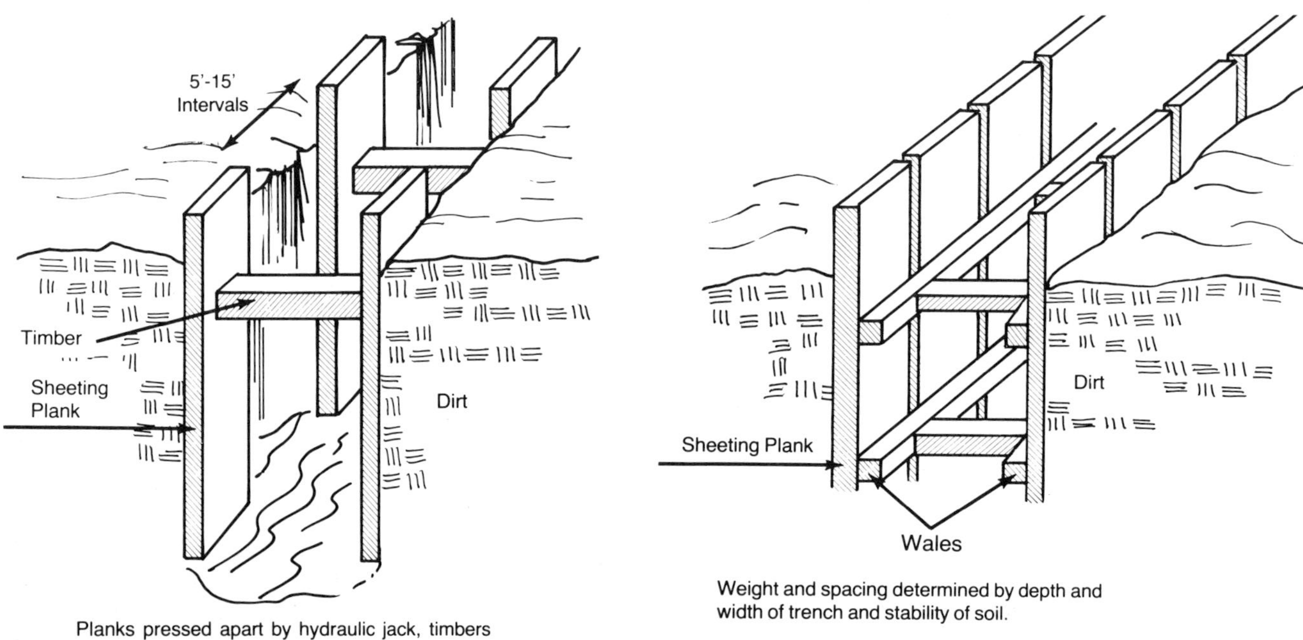

Figure 4.6 Shoring for trench where there is a slight danger of caving.

Figure 4.7 Shoring for trench where danger of caving is greater.

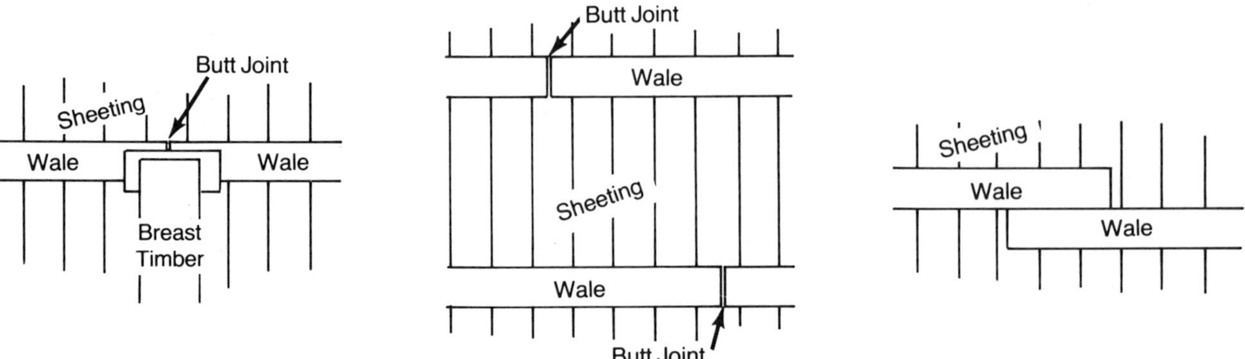

Figure 4.8 Details of bracing for trench shoring. Depending on conditions, sheeting should be long enough to reach one or two feet below bottom of trench and one or two feet above the top of trench. Wales should be no farther apart vertically than five feet and should have two or more breast timbers spaced five or more feet apart.

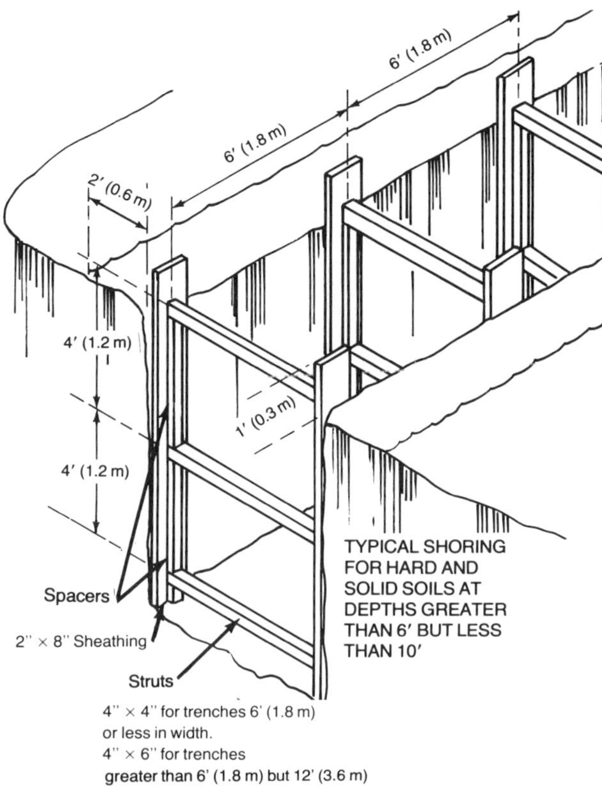

Figure 4.9 Typical shoring for hard and solid soils at depths greater than six feet but less than ten feet. *Courtesy of Construction Safety Associations of Ontario.*

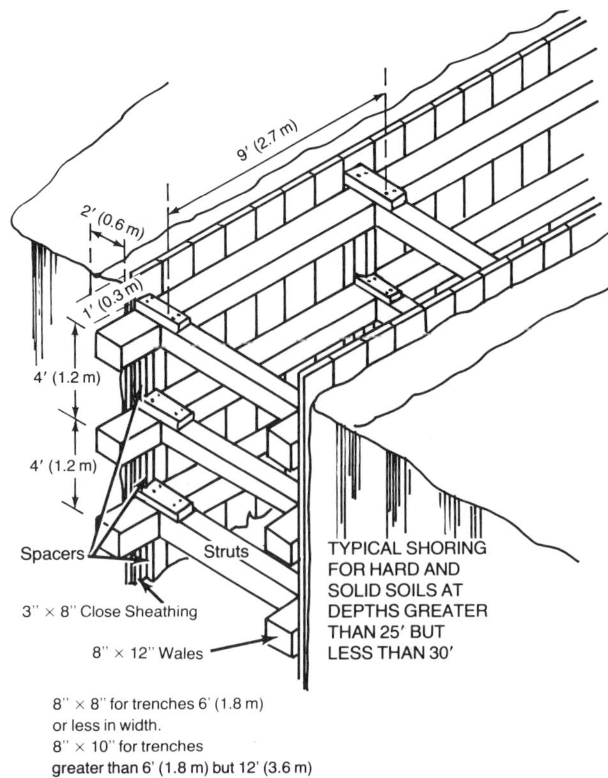

Figure 4.10 Typical shoring for hard and solid soils at depths greater than 25 feet but less than 30 feet. *Courtesy of Construction Safety Associations of Ontario.*

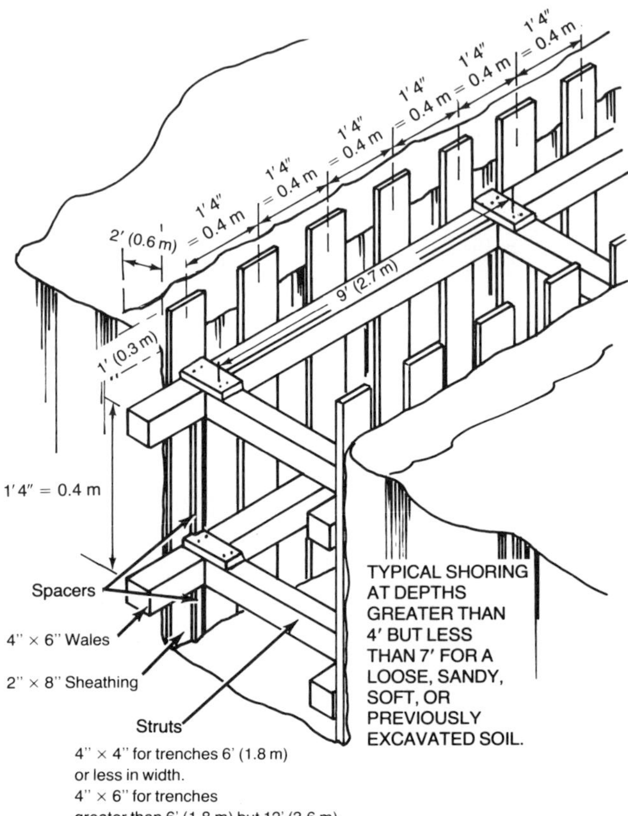

Figure 4.11 Typical shoring at depths greater than four feet but less than seven feet for a loose, sandy, soft, or previously excavated soil. *Courtesy of Construction Safety Associations of Ontario.*

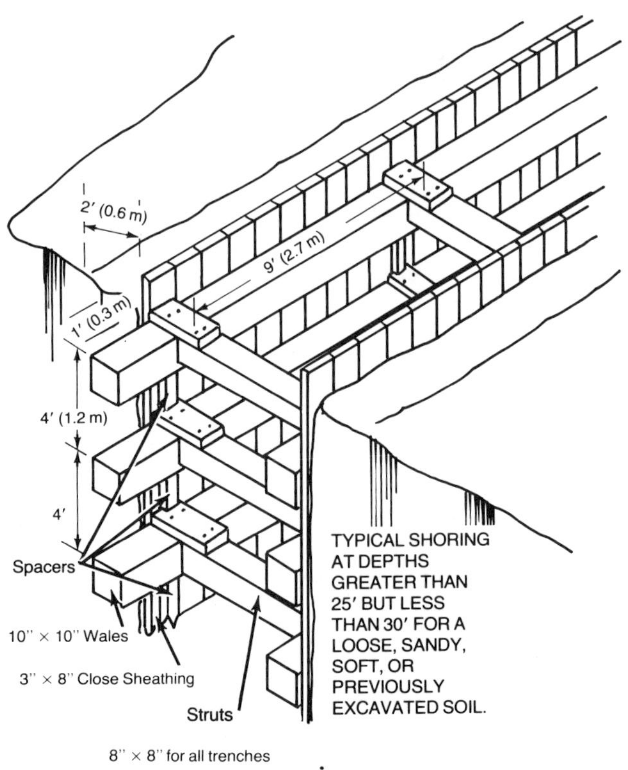

Figure 4.12 Typical shoring at depths greater than 25 feet but less than 30 feet for a loose, sandy, soft, or previously excavated soil. *Courtesy of Construction Safety Associations of Ontario.*

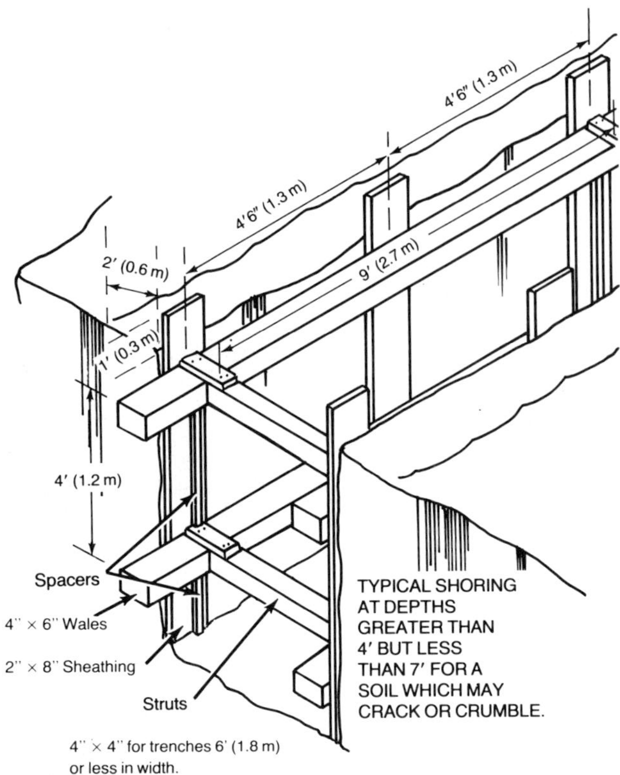

Figure 4.13 Typical shoring at depths greater than four feet but less than seven feet for a soil that might crack or crumble. *Courtesy of Construction Safety Associations of Ontario.*

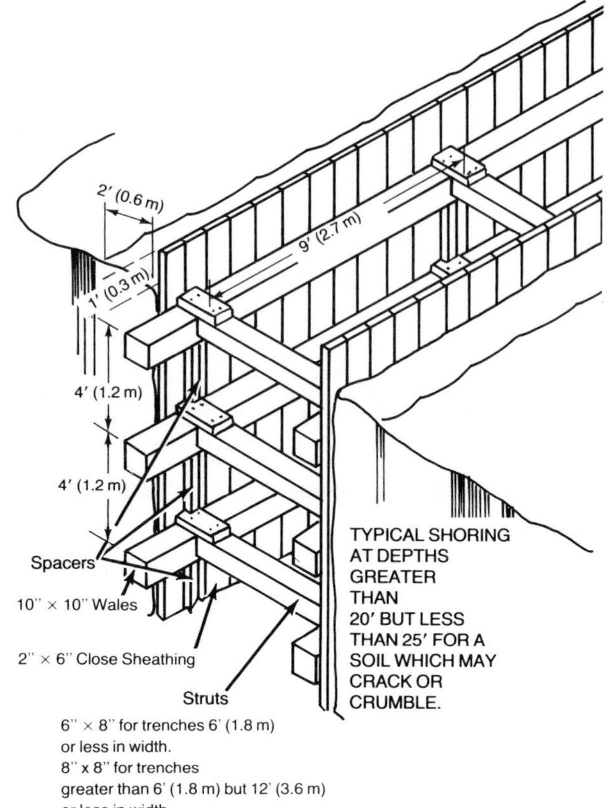

Figure 4.14 Typical shoring at depths greater than 20 feet but less than 25 feet for a soil that might crack or crumble. *Courtesy of Construction Safety Associations of Ontario.*

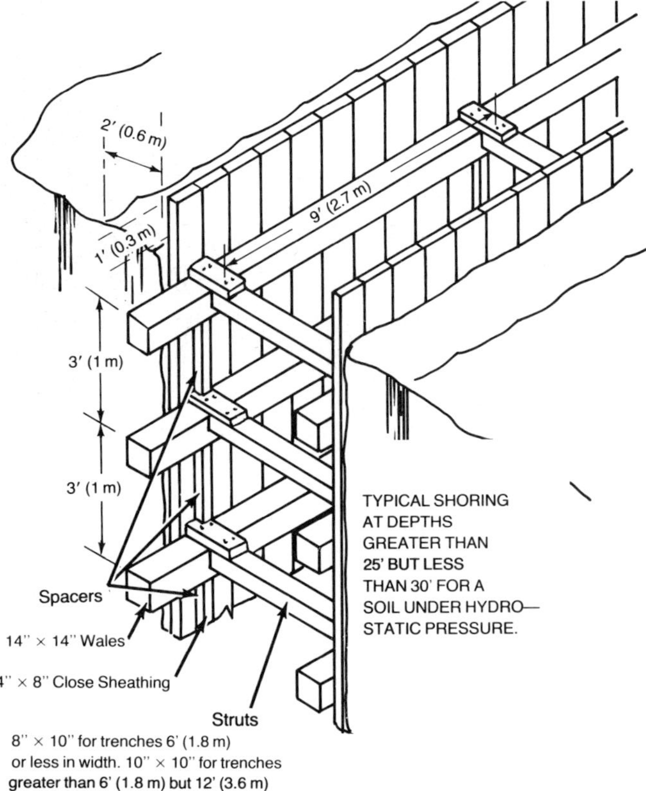

Figure 4.15 Typical shoring at depths greater than 25 feet but less than 30 feet for a soil under hydrostatic pressure. *Courtesy of Construction Safety Associations of Ontario.*

Figure 4.16 Trenches may be shored with materials at hand, such as debris and spineboards, if necessary. *Courtesy of Montgomery County, Md., Fire/Rescue Services.*

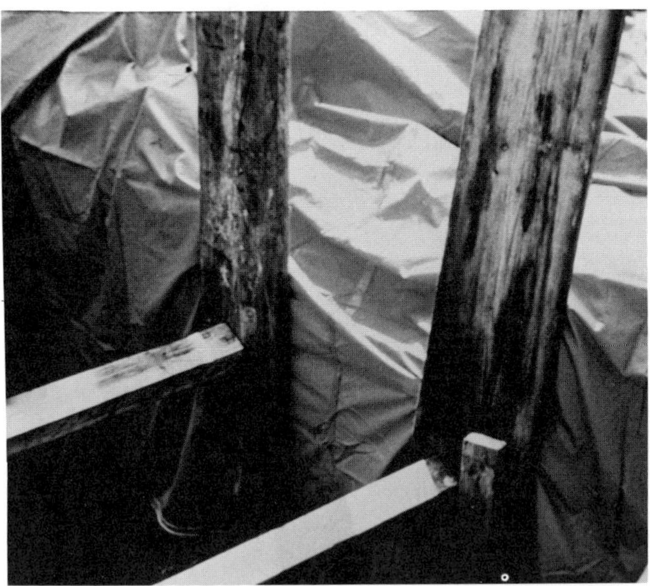

Figure 4.17 In relatively firm ground a salvage cover can be used for shoring. *Courtesy of Montgomery County, Md., Fire/Rescue Services.*

Figure 4.18 The long side of a trench jack foot should run parallel with the long side of the timber it braces. *Courtesy of Montgomery County, Md., Fire/Rescue Services.*

Figure 4.19 Be careful not to overextend trench jacks. *Courtesy of Montgomery County, Md., Fire/Rescue Services.*

Figure 4.20 Nailing should be kept to a minimum, but struts should be toenailed to prevent their being bumped out of place. *Courtesy of Montgomery County, Md., Fire/Rescue Services.*

BREACHING WALLS

Many different types of construction will be encountered in rescue operations, including walls made of brick with lime mortar, brick with cement mortar, stone, concrete, and concrete block.

When cutting through or breaching walls or floors of large buildings, try to locate sections of the structures in which cutting can be done most quickly and safely. When cutting through walls, be sure not to weaken support beams and columns. After a building has been subjected to severe stress, as is likely during intense fire or explosions, the parts left standing may appear sound, although badly shaken and cracked. Therefore, when breaching wall sections, especially with air hammers, take care to prevent further collapse (Figure 4.21).

Figure 4.21 Choose carefully the place for breaching walls to prevent further collapse. Here firefighters are breaking through a bricked-up doorway, knowing that it is not a load-bearing spot. *Courtesy of Montgomery County, Md., Fire/Rescue Services.*

Openings large enough for rescue purposes can usually be made in brick walls without danger of the masonry falling. The bricks should be removed so the opening is shaped like an arch.

In all but concrete walls and floors, the best method is to cut a small hole and then enlarge it. With concrete, however, it is better to cut around the edge of the section to be removed. If the concrete is reinforced, the reinforcing bars can then be cut with a hacksaw or a torch and the material removed in one piece. If a torch is used, be sure explosive gases are not present and that flammable materials are not ignited. A charged line should be kept ready nearby.

SHORING

Shoring is erecting a series of timbers or jacks to strengthen a wall or to prevent further collapse of a building or earth opening, not to restore walls or floors to their original position. Any attempt to force beams, floor sections, or walls back into place may cause further collapse and damage. It is important, however, to secure all shores in position. This action must be done gradually, without shock. Since bracing, shoring, and supporting walls and floors are basically engineering problems, fire officers should call upon the sources available for assistance when planning and conducting these operations. Shoring is difficult work and requires training and practice before efficiency is attained.

Raking Shore

If a wall is bulging or out of plumb, shores can be used to brace the wall or hold it in position, especially if excavating or tunneling is being done next to it. These shores are called bracing, pushing, or raking shores (see Figures 4.22 and 4.23). The principal parts include wall plate, raker, and sole plate.

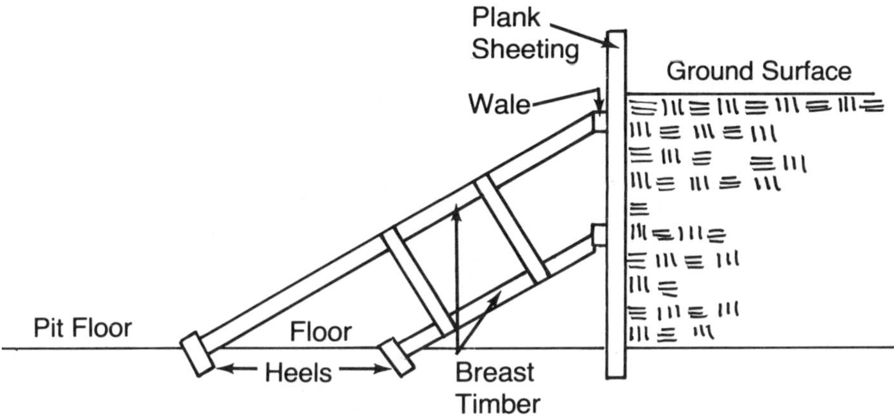

Figure 4.22 Side view of basic raking shore.

Wales = 6" × 6" (15 × 15 cm), minimum, vertical spacing not less than 5' (1.5 m)

Sheeting Plank = 12" × 13" (30 × 33 cm), minimum, extends at least 1' (30 cm) below trench floor and 1' (30 cm) above surface

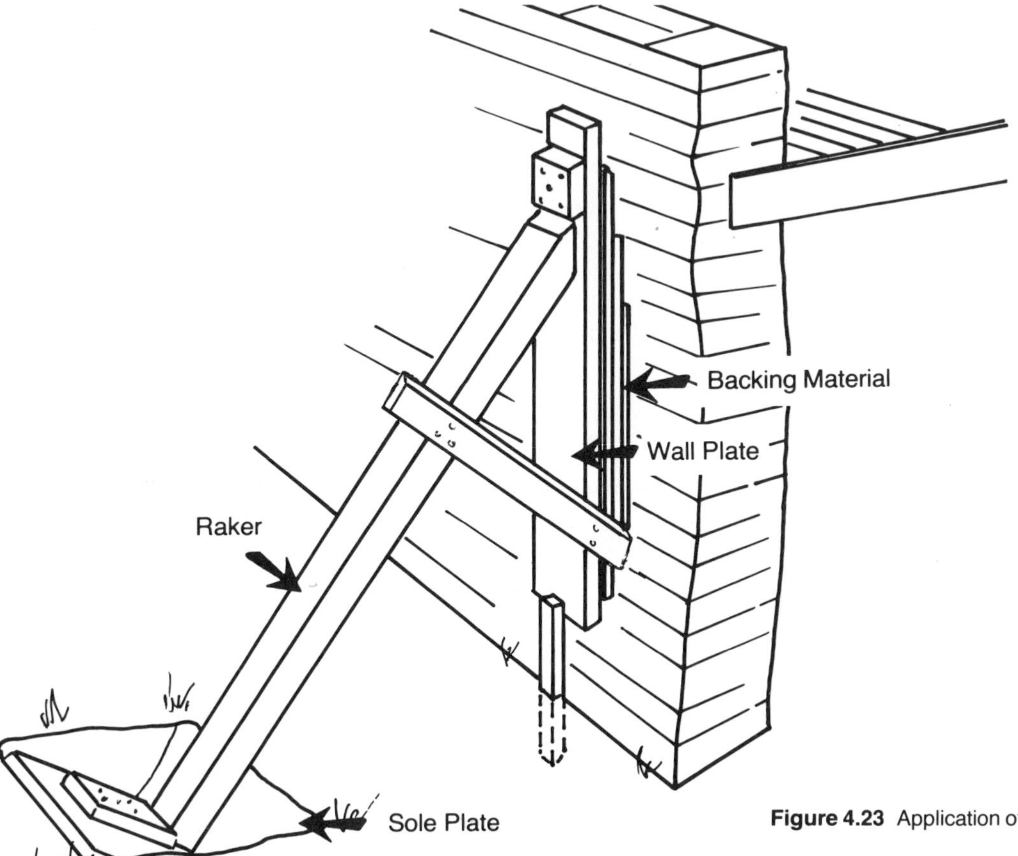

Figure 4.23 Application of raking shore to unstable wall.

If possible, the wall plate should extend the length of the wall. When used against a bulging wall, back the wall plate with timbers to provide continuous pressure.

Rakers are best formed with square timbers. The number required varies from one to four, depending on the height of the wall to be supported and the number of floors carried by the wall. Table 4.1 is a rough guide to the number of rakers and the size of timber required for shores of different heights.

TABLE 4.1
Number and Size of Rakers Needed

Height of Shore		Rakers	Raker Area (square section)	
Feet	Meters	Needed	in ^{2}	cm ^{2}
20-30	6-9	1	25	162
30-40	9-12	2	36	234
40-50	12-15	3	50	325
50+	15+	4	80	520

Flying Shore

A flying shore is used to brace a damaged wall when a sound adjacent wall can be used as a support. Two types are shown in Figure 4.24. Principal parts are horizontal beams, wall plates, and struts. Other items are cleats, wedges, and straining pieces.

To erect a flying shore, nail the cleats on the wall plate, one pair to support the horizontal beam or shore and the others to support the struts. Set the struts at an angle not greater than 45 degrees to the horizontal beam and keep them apart with straining pieces. The length of the straining pieces is determined by the length of the horizontal beam.

Proper measurements and angles can best be determined by laying the job out on the ground before erection. While holding the wall plates in position, place the horizontal beam on the center cleats and tighten with wedges and shims inserted between the shore and wall plates. Next, place struts and straining pieces into position. Cleats may be used to brace the shore more rigidly.

The wall plate should be continuous throughout its length, with packing between the wall and wall plate, if necessary, to provide continuous pressure.

Flying shores should be placed along a wall at intervals of 8 to 12 feet (2.4 to 3.6 m), depending on the situation, type of wall, and the degree of damage. Do not use straight flying shores between two walls separated by more than 25 feet (7.6 m).

A weakened foundation or damage to the lower portion of the wall often makes it unstable. Causes of instability in a structure subjected to blast may vary. Since there is no standard method of approach, meeting situations that require shoring and supporting walls and floors calls for careful planning and good workmanship. The lower part of the wall and its footing or foundation must carry the entire weight of the structure above it. If damaged by blast or by removal of an adjacent supporting structure, the wall may buckle or crumble. Therefore, bracing or shoring on lower parts of the wall should be stronger than corresponding work on the upper portions.

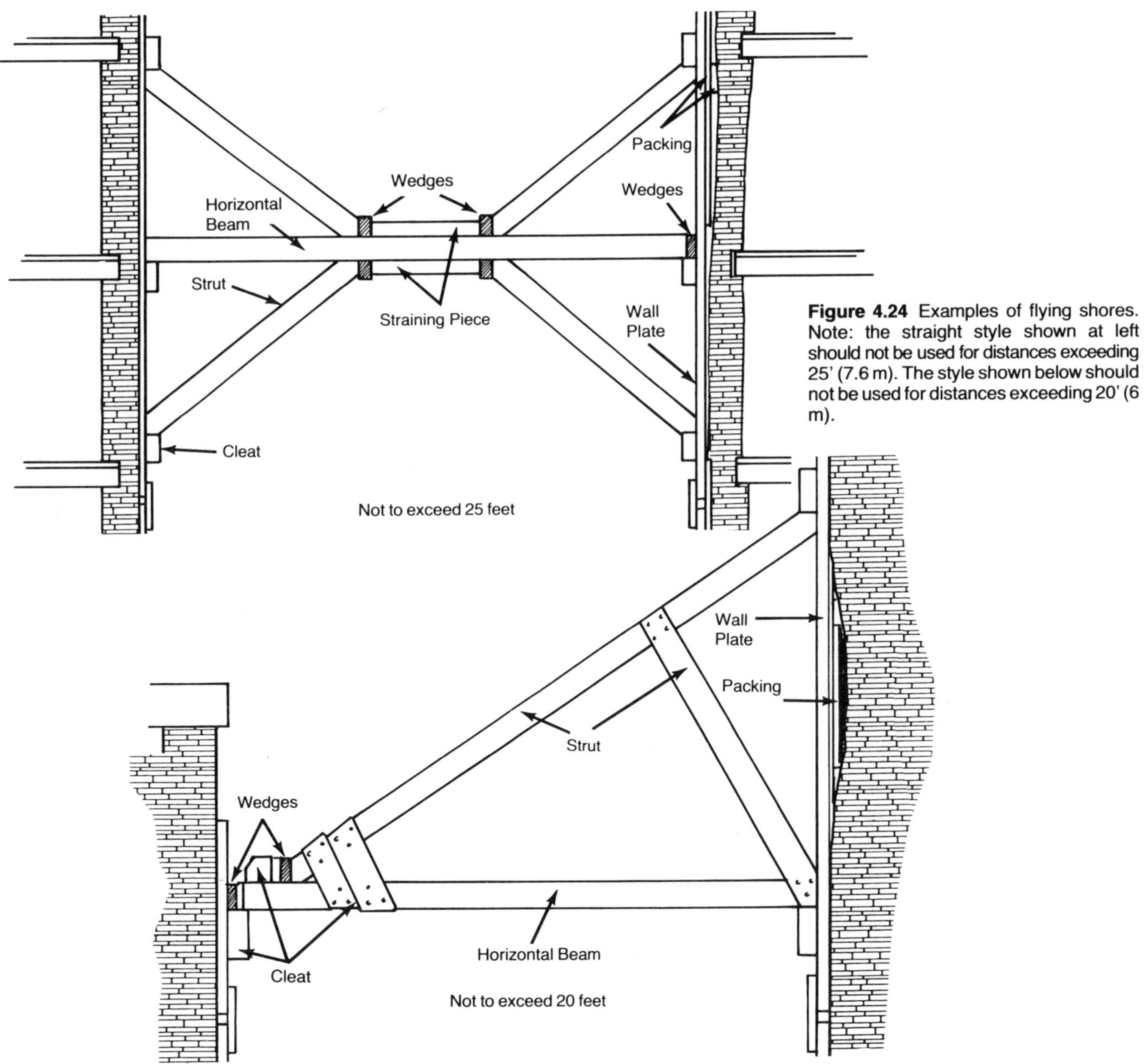

Figure 4.24 Examples of flying shores. Note: the straight style shown at left should not be used for distances exceeding 25' (7.6 m). The style shown below should not be used for distances exceeding 20' (6 m).

Dead Or Vertical Shore

A dead or vertical shore is used to carry the vertical dead load of a wall or floor (Figure 4.25). The principal parts are the strut, the sole plate, and the headpiece.

Struts should be made of square timber heavy enough to carry the maximum expected load. It is difficult to estimate what load a strut must carry and to gauge what load the strut timber can support. However, in strutting a damaged building the following principles apply:

- For a given size of timber, the shorter the strut, the greater load it can carry.
- A strut of square cross-section is stronger than a rectangular one of the same area.
- A strut will be much stronger in service if its ends are cleanly cut to fit squarely against the sole plate and headpiece.
- Struts should always be made a little heavier than appears necessary. The size should be determined by the height and weight of the wall or floor to be supported.

Set folding wedges under the strut and drive the strut into position with the wedges until it just takes the weight and no more. Wedges should not be driven tighter since that would lift the wall or floor being supported and might cause more damage to the building.

Figure 4.25 An example of a dead (vertical) shore.

The sole plate should be made as long and as wide as practicable to spread the load over a sizable area. The sole plate should not be placed on a cellar arch or timber floor if there is doubt that the arch or floor can carry the load. In such cases, the sole plate must be supported from below. Where struts are used on the upper floors of buildings, the strutting should be repeated on all the floors so the load will have a solid foundation. An exception to this is when a strut can be supported on a heavy beam in a part of a building that has not suffered much damage.

The headpiece should have about the same cross-section as the strut. However, the load being carried will be a determining factor here; so will the span between struts where two struts are used. This span should be kept as short as possible because the smaller the span, the greater the load the headpiece can carry.

Cribbing

Cribbing has various uses

Cribbing is arranging planks into a crate-like construction with separated joints. These planks may sometimes be criss-crossed with the joint staggered. Cribbing in rescue work is mostly used to support roofs and ceilings, but it can be used on walls when necessary. Tunnel openings and passages can be secured by shoring and cribbing.

RESCUE FROM HAZARDOUS ATMOSPHERES

Fire departments are often called to rescue persons from situations involving hazardous atmospheres. These situations warrant close examination by the firefighter.

Hazardous atmospheres may be grouped into three types:

- Atmospheres containing flammable vapors or gases in amounts that are combustible or explosive.
- Atmospheres containing toxic or corrosive vapors that may poison or injure the rescue worker.
- Atmospheres in which the oxygen content has been diluted or depleted until the atmosphere has too little oxygen to support life.

For an atmosphere to be hazardous, it need only meet the criteria for one of these situations, but often the atmosphere will meet the criteria for all three. No attempt will be made in this manual to discuss or describe the individual gases or vapors. However, the hazards resulting from these atmospheres must be taken into account.

There are two basic rules for firefighters who must work in or around hazardous atmospheres:

- Get as much information about the situation as possible, and react accordingly.

- When sufficient information has not been obtained, assume the situation to be the worst possible.

Size up and SCBAs necessary

Victims trapped in a hazardous atmosphere are no better off if firefighters rush hastily to their aid and are overcome by the same hazard. The would-be rescuers must take time to size up the hazard and then take the necessary precautions before entering the area. Entry should not be attempted if the victim is obviously dead unless no risk to the firefighter is involved.

Sources of information at an emergency scene are numerous. Generally, there is adequate information to prevent firefighters from falling victim to the hazard. There are four areas firefighters can rely on for information:

1. What they already know from:

   Pre-fire planning

   Occupancy use

   Previous calls to that particular occupancy

   Dispatch information

2. What they find out by questioning:

   Owners or workers

   Guards or police

   Victims

   Witnesses

3. What they observe (see, smell, or hear) about:

   Physical conditions —

   - climate
   - fire
   - smoke
   - smells
   - sounds
   - container shape
   - victim symptoms

   Shipping papers —

   - bills of lading (trucks)
   - waybills (trains)
   - consists (trains)
   - cargo manifests (ships)

Marking Systems —

- DOT
- United Nations
- Manufacturer labels

4. Information they can obtain from:

    Hazardous material guides

    Manufacturers

    Distributors

    Plant safety personnel

    Trade associations

    CHEMTREC

    CANUTEC

After adequate information is gathered and entry is to be attempted, proper protective equipment must be worn. This includes complete body protection and breathing apparatus. Any breathing apparatus used in a hazardous atmosphere should be positive pressure. If demand apparatus is to be used, the bypass could be cracked open slightly to provide a positive pressure to prevent gases from entering the mask during the slight negative pressure caused by inhalation.

Radio communication is helpful, but should be used only if the radio is listed and approved for use in explosive atmospheres. Any electric hand tool or appliance to be used in flammable atmospheres must also be listed and approved for use in explosive atmospheres. Electric power to an area involving a flammable atmosphere should be cut from outside the area. This may involve contacting the power company to cut the power to an area. Generally, though, it can be done with breakers on the premises.

Ventilation may help

When rescuing persons from an atmosphere in which there is toxic gas, ventilation should be considered. Ventilation will reduce the effect of the gas on the victims and will lessen the rescuers' risks. Self-contained breathing apparatus is a necessity.

Hazardous atmospheres resulting from flammable or combustible liquid spills require additional ventilation consideration. Generally, as long as the spill remains it will provide flammable vapors. Therefore, the spill must be contained before ventilation can be considered. Applying foams in combination with ventilation is recommended in these situations. However, foam is not completely effective in all circumstances, and firefighters should still be cautious. The atmosphere should be monitored with a combustible-gas indicator to evaluate the risk to rescue workers (Figure 4.26).

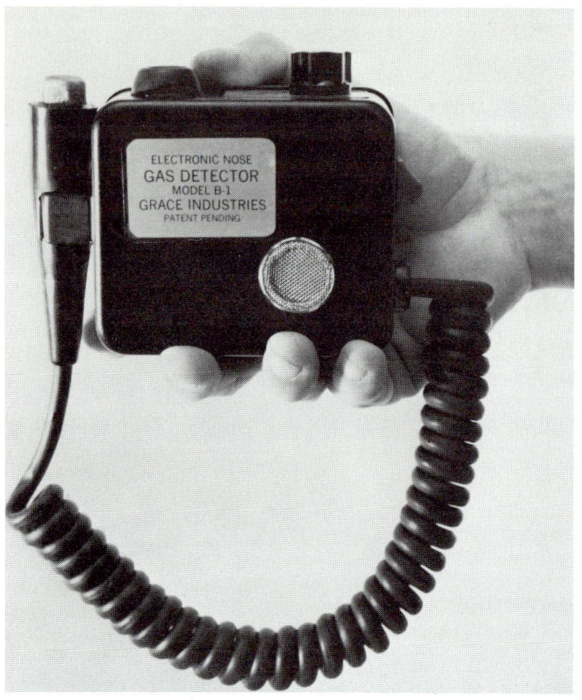

Courtesy of Grace Industries.

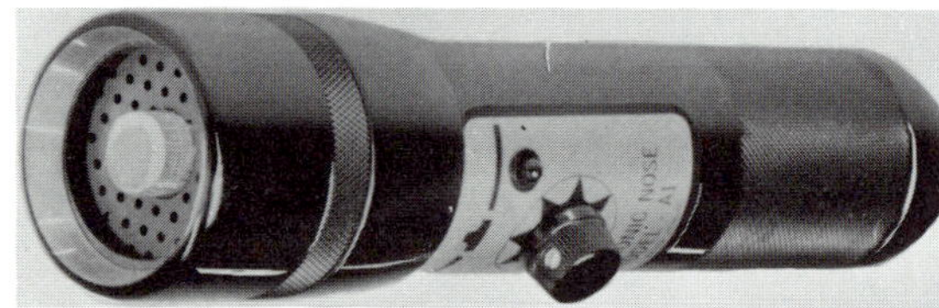

Courtesy of Grace Industries.

Courtesy of Scott Aviation

Figure 4.26 Combustible-gas detectors are made in different forms for various conditions and uses.

RADIATION ACCIDENTS

Firefighters must be alert to the dangers of radiation accidents because the use of radioactive materials at atomic energy installations, industrial labs, hospitals, nuclear power plants, and educational institutions has increased dramatically. Firefighters may at any time be called to rescue victims exposed to one of these radiation sources. Often the incident will involve only a few people, but the danger to rescuers will be high if they are not properly trained.

Trained rescuers should not be afraid to handle cases of radiation exposure or radioactive contamination. It is usually a matter of common sense, cleanliness, and planning. Although radiation cannot be detected by the senses, it can be detected and measured by special instruments.

Wear full protective clothing

If any amount of radiation is present, no matter how small, no rescue attempt should be made without full protective clothing. Wear self-contained breathing apparatus and full turnout gear, including turnout pants, gloves, boots, and helmet.

Even if well equipped, rescuers should not jeopardize themselves attempting to rescue a victim from a radiation hot-zone that has been declared to be above lethal dosage. Of course, the expert on the scene who makes such a declaration has to be sure. It is a thankless task, but someone has to decide who will be treated first in a large disaster involving many injured persons. Often it will be a firefighter.

Types of radiation victims

There are four types of radiation accident victims. Victims with a lethal dose of external radiation are no hazard to rescuers. If they receive only a smaller dose and can be saved, so much the better. They are no more dangerous to rescuers than a diagnostic X-ray patient.

Another type is the individual who has received internal contamination by inhalation or ingestion. Such victims also are no hazard to attendants, other patients, or the environment. After being cleaned of minor amounts of contaminated material deposited on the body surface during airborne exposure, they are similar to chemical poisoning cases.

External contamination of body surface and/or clothing by liquids or dirt particles presents a third type, with problems similar to vermin infestation.

When external contamination is complicated by a wound, take care not to cross-contaminate surrounding surfaces from the wound and vice versa. For this fourth type of patient, clean the wound and surrounding surfaces separately and seal.

Get help from experts

Rescue personnel are usually the first members of the medical team to see cases of radiation exposure or radioactive contamination. Their first acts will depend on whether they evacuate victims from a nuclear energy plant, from a university or medical group regularly working with nuclear material, or from a road transportation accident. Trained, knowledgeable co-workers, supervisors, or hygienists are usually on hand at the plant but not at the road site.

When the accident occurs at a plant, hygienists, supervisors, co-workers, and victims should be able to tell rescue squad members about the accident, the number of victims, the type of radiation exposure or radioactive contamination involved, and body areas possibly affected. A gross measurement of the amount of radiation involved can be helpful.

Responsibilities of the Rescuer To the Victim

- Give lifesaving emergency assistance if needed.
- Secure pertinent information, including an estimate of radiation, from those present.

- Determine if physical injury or an open wound is involved. Cover the wound with a clean dressing held in place with an elastic bandage. Do not use adhesive.
- Cover the stretcher, including the pillow, with an open blanket, and wrap the victim in the blanket to limit the spread of contamination.
- Give the hospital all available information by radio or telephone.

To Rescue Personnel

- Survey clothing, ambulance, and other equipment on arrival at the hospital before undertaking further activity.
- Dispose of contaminated clothing in a container marked "Radioactive—Do Not Discard." Wash or shower with cool water.
- Survey rescue squad personnel in the contaminated area with a radiation-survey meter and record the measurements. Cleansing must continue until a responsible physician indicates the person may leave.

Radiation Detection Devices

None of their senses will warn people of the presence of nuclear radiation. Instead, they must rely on some secondary means of detection. The term *detection* in this manual includes even the indication of the presence of nuclear radiation. The roentgen is a unit of radiation exposure. Since this unit is rather large for measuring common exposures, the milliroentgen, or 0.001 roentgen, is frequently used for measuring small exposures.

Figure 4.27 A pocket dosimeter used to detect radiation.

A dosimeter is an instrument that will measure gamma radiation exposure but not alpha or beta radiation. Rescue workers sometimes need an instrument such as the dosimeter, which is capable of measuring an individual's total exposure to gamma radiation (Figure 4.27). If the dosimeter reading is zero at the start of the exposure period, the exposure may be read directly from the dosimeter. If a dosimeter has previously been used, any initial reading must be subtracted from the final reading for a correct indication of the exposure. Dosimeters should be used when personnel are likely to be accidentally exposed to doses of radiation. They are adequate detecting devices for short periods of high radiation or long periods of low radiation exposure. The dosimeter will also register small repeated doses of radiation over a long period.

Survey or rate meters measure exposure rates in roentgens or milliroentgens per hour (Figure 4.28). Survey operations require an instrument that will measure exposure rate rather than the total roentgens.

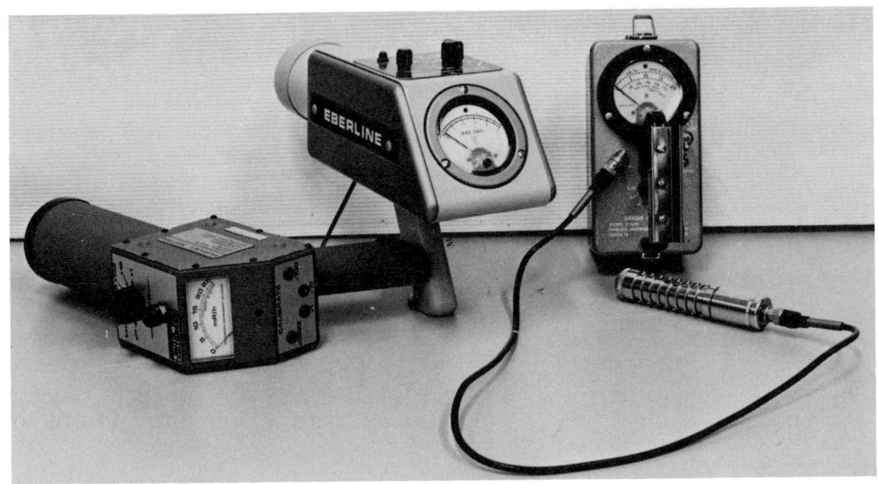

Figure 4.28 Rate meters used to measure roentgens or milliroentgens per hour.

RESCUE FROM ELECTRICAL CONTACT

When live electrical wires are knocked to the ground because of wind storms, lightning, fires, or other reasons, the fire department's main function is to guard the wires and call for the power company to remedy the situation. However, if a firefighter finds a victim in contact with live wires or energized electrical equipment, the correct rescue procedures must be known and implemented immediately. Firefighters should not proceed with such haste that they forget their own safety. Injured rescuers certainly reduce the chance of a successful rescue.

Lineman's rubber gloves (Figure 4.29) and leather protectors are a must for firefighters anytime they are working close to energized electrical equipment and wires. However, these gloves and leather protectors are not made for handling live wires or equipment. These gloves only give firefighters added protection when using energized-wire cutters and other nonconductive tools to rescue a victim who is in contact with energized wires or equipment.

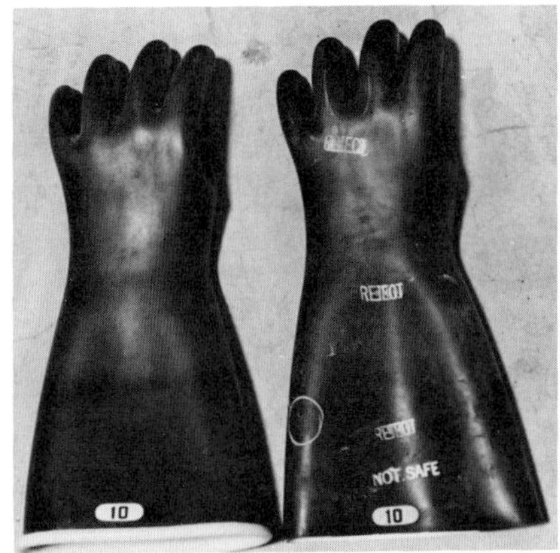

Figure 4.29 If electrical wires must be handled, lineman's gloves are required. Caution: Gloves must be tested regularly to assure retention of protective qualities. The glove on the right did not pass the test so it is marked "reject" and "not safe."

Lineman's rubber gloves and leather protectors are made of special rubber that has been electrically tested with special equipment. These gloves must be carefully inspected by qualified personnel after each use. Consult the local power company for assistance. After the gloves have been inspected, they should be stored in a special bag away from oil, grease, and sunlight. Contact with any of these may damage the gloves. The lineman's rubber gloves and protectors should be used only for electrical protection because even a pinhole in one of the gloves will eliminate their safety factor. In addition to being tested after each use, the rubber gloves must also be electrically tested at a testing lab twice a year, or more often if damage is suspected. Again, consult the local power company for help in locating a testing laboratory.

Firefighters should remember not to rely on their rubber boots for protection against electric shock. Boots may contain enough conductive carbon black to transmit a dangerous shock.

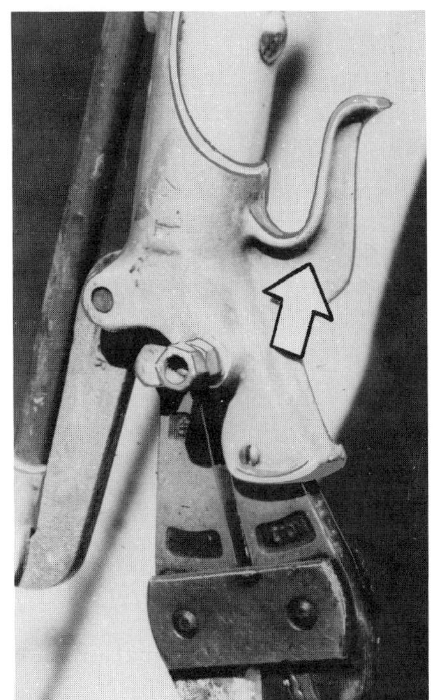

Work as fast as possible to remove a victim, and avoid touching the victim or the electrical equipment with any tool that may conduct electricity. Nonconductive rescue equipment includes a special insulated hot-stick cutter (Figure 4.30), dry rope, or dry wooden ladders or poles. Any moisture on or in the equipment will conduct electrical current. Even with dry equipment, firefighters removing wires from a victim should use tools providing the maximum possible distance from the electrical source.

If a live wire is lying on a victim, it should be moved by the safest and fastest means possible. A quick, single, well-placed motion is necessary because any slow movement may cause contacts and injuries from the wire. Safer control of such a wire comes from using a nonconducting tool to *pull* the wire toward the rescuer while stepping away. Secure the wire after removal with some nonconducting material, such as a wooden ladder, so the wire does not move and jump around dangerously.

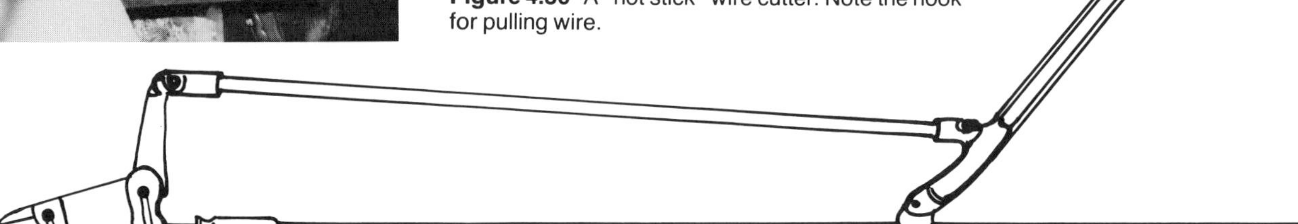

Figure 4.30 A "hot stick" wire cutter. Note the hook for pulling wire.

If the victim is lying on a live wire and is not entangled in it, or if the victim is touching energized equipment, the best method of rescue is still to use a nonconducting hot stick or dry rope to pull the victim clear. A power company type of insulated hot stick is best. With this maneuver it is necessary to secure the wire while pulling the victim so the wire will not move or whip around. Anchor the wire to the ground with a heavy, nonconductive object such as a piece of wood or pole. Less electricity should flow to the victim if the wire is held in contact with the ground. A quick movement should not be used to free a victim. Although other burns may be received during a slower movement, the alternative of leaving the victim in contact with the live wire is much worse.

Cutting live wires may be necessary if the victim is entangled in them. This is a dangerous operation that untrained firefighters or firefighters without the proper equipment should never attempt.

When wire is manufactured it is coiled on reels, and it often retains some of this original coil after it has been installed. If a wire is cut, it may curl and roll along the ground. Always secure a wire before cutting it so the loose ends will stay under control.

One suggested method is to throw two objects across the wire to pin it to the ground. Cut the wires on both sides of the victim with a suitable approved hot-line wire cutter. The wire must be cut on both sides of a victim and then separated from the ends that are secured by the two objects. Hot-line cutters are mounted on insulated sticks, and they all have their safety limitations. They are not bolt cutters with insulation on the handles.

Cutting live wires can cause other hazards. A loss of tension could cause other wires to sag to the ground some distance from where they are cut. With any wire, it may be impossible to determine from its appearance if it is energized. Even live it may not be arcing. Also, automatic switching equipment may reenergize fallen wires. Wires must be cut on both sides because they are frequently energized from both directions. Single-cut wires may be in contact with other live wires elsewhere.

Consider any wire live

Nevertheless, wires may at times need cutting to minimize the hazard to a victim, to firefighters, or to property. Sometimes, in order to gain access to a fire area, wires must be cut before the power company arrives. Caution must be used, and cutting must be done only in accordance with practices agreed upon beforehand during discussions with power company safety officials.

During an electrical rescue operation, additional personnel are required for crowd control. Curious persons attracted by the sirens, red lights, and activity may not see or recognize the hazard. Whipping, arcing wires can move along the ground and cause injury. Less obvious is the possibility that broken or cut wires will cause increased stress on the supporting pole or tower on either side of the break. Failure of such poles or towers over a crowd of observers could cause much injury.

Rescued victims will occasionally be in cardiac arrest because of the paralyzing effect of the current. Paralysis of the breathing center of the brain may also occur. In these cases, immediate administration of cardiopulmonary resuscitation is a must. See IFSTA's **Fire Service First Aid Practices** for complete details of this life-saving technique.

Search Guidelines

Chapter 5

NFPA STANDARD 1001
RESCUE
FIRE FIGHTER I

3-12.1 The fire fighter shall demonstrate the removal of injured persons from the immediate hazard by the use of carries, drags, and stretchers.

3-12.2 The fire fighter shall demonstrate the procedure for searching for victims in burning, smoke-filled buildings, or other hostile environments.

3-15.2 The fire fighter shall demonstrate techniques for action when trapped or disoriented in a fire situation or in a hostile environment.

Reprinted by permission from NFPA Standard No. 1001, *Standard for Fire Fighter Professional Qualifications.* Copyright © 1981, National Fire Protection Association, Boston, MA.

Chapter 5
Search Guidelines

When searching for victims in a fire, rescuers must always be concerned with safety. Personnel should be properly trained and equipped with the necessary tools to accomplish the rescue in a minimum of time. Unsafe, hurried rescue attempts could be fatal to rescuers and victims.

SAFETY IN BUILDING SEARCH

Every time the firefighter responds to a fire, a human life may be in jeopardy. In those instances where life is endangered, a search must be initiated rapidly. However, to insure that these strenuous situations do not become disastrous ones, rescuers must operate safely with sound judgment.

Keep safety in mind while searching

Following is a list of common-sense safety points of which rescuers should be aware before they attempt any type of search and rescue within a building:

- If fire conditions are so advanced or the condition of the building is so poor that rescuers have a good chance of losing their lives, rescue should not be attempted. Under such conditions it is unlikely the victim would be alive.

- When a backdraft is possible, attempt entry only after ventilation is begun. Entry before proper ventilation could result in a backdraft explosion, often causing serious injury (Figure 5.1).

- Always wear full protective gear (and wear it properly). This includes self-contained breathing apparatus.

- Always work in pairs and keep in constant contact, remembering that each is responsible for the other.

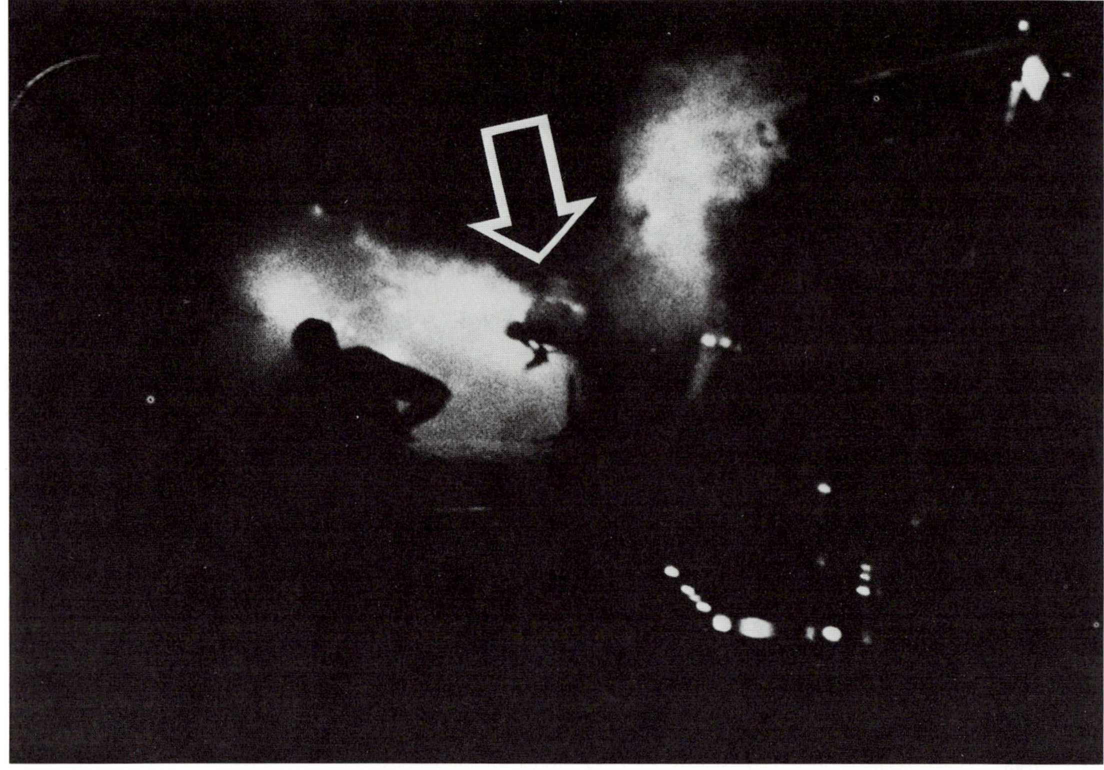

Figure 5.1 A backdraft can injure hasty rescuers. *Courtesy of Nancy Engebretson.*

Search systematically

- Have a plan or objective. Do not wander aimlessly. Working systematically will reduce the possibility of disorientation.
- Make sure that there is a secondary means of egress available for all firefighters involved in the search.
- When operating on the floor above the fire, have a charged line ready. Hoselines can be used as lifelines in addition to supplying emergency fire fighting capability.
- Mark entrance into rooms and make note of the direction turned while going into the room. To exit and return, turn in the opposite direction.
- Feel doors for excessive heat before opening.
- Stay low and move cautiously.
- Stay alert (use all senses).
- Watch for hot spots and weakened structures.
- Keep in contact with a wall (ropes or straps can extend coverage).
- As the rescuer moves around a room, windows may be opened to relieve heat and smoke if such venting does not extend the fire.

- If fire is encountered, closing a doorway will often contain the fire temporarily, allowing the searchers to continue.
- Once the search is complete, searchers should report back promptly to the officer in charge.
- Trapped rescuers can obtain some protection by hiding behind a solid, well-hung door.
- When trapped, firefighters as a last resort should throw their helmet outside to the ground. A helmet thrown to the ground outside a fire building indicates that a firefighter is trapped inside.

Throw helmet as last resort

Stay calm if you lose your direction. Follow the wall and it will bring you to the door through which you entered or a door to another area. If you can come across a hoseline, crawl along this line. It will take you to the nozzle team or lead you outside. If you become trapped, follow the wall to the nearest window and signal for assistance.

Always carry a good handlight. You can use it to signal your location. If you feel you are losing consciousness, set the handlight on the floor with the light shining on the ceiling.

Open doors carefully. Feel the door first to judge the heat behind it. Don't stand in front of the door. Stay to one side, keep low, and open the door. If there is fire behind the door, this will allow the heat and combustion products to pass over your head.

If a door is hard to open inward, do not kick the door open. A victim may have collapsed at the door while trying to escape. Slowly push the door open, feeling behind it to check for the possible victim.

THE BUILDING SEARCH

Searching a building has two objectives: finding victims and obtaining information on the extent of fire.

Upon arrival at the fire scene, check with building occupants who have escaped the fire for information about those who might still be inside. If others are inside the building, ask those who escaped where the others might be. Make sure the information is factual. Neighbors might know of the occupants' possible location, because neighbors can be familiar with habits and room location. Also, the victim might have been seen at the window just before your arrival.

Rescue teams should always make their entry into the building known to their company officer. In no case should rescuers enter a building without first being sure that a responsible person outside the structure knows what they are doing.

BUILDING SIZE UP

It takes only a few seconds to stop and look at a building before entering and to gather facts that will be useful to you when you go inside. Look at the entire building you are about to enter and at its surroundings. Then when you get to a window inside you can figure out your location.

ESTABLISHING A SEARCH PATTERN

Search in pairs

The buddy system of searching for victims should always be used. Two persons using an efficient system can search an average residence in a short time.

The fire floor should be checked first, and the floor directly above the fire should be checked next (or at the same time if enough personnel are available). Rescuers on the floor above the fire should have a charged hoseline with them. Your initial size up should have told you if there could be an attic apartment. If the roof is flat, there is no need in most residences to search for a staircase leading from the top floor.

Exit from same place entered

When multiple rooms or apartments lead off a center hallway, the rescuers' search will be a series of room or apartment searches. When entering the first room, the pair of rescuers will turn right or left, thereby establishing a search pattern that may be used to search any building, from a one-story, single-family home to a large high-rise building. Upon exiting the room, the rescuers should turn in the same direction they entered and continue their search for another room. It is important that rescuers exit a room or apartment where they entered to insure a complete search.

Rooms that have been searched should be marked to avoid duplication of effort (Figure 5.2). Several methods of marking rooms searched are used by the fire service—furniture in doorways, chalk marks, and latch straps (see page 87) over the door knobs. Standard operating procedures may dictate any method of marking; whichever is used must be clearly known and understood by all personnel who may participate in the search.

If rescuers have to abort their search or if a victim is located and is to be removed from the building, rescuers should leave by turning in the opposite direction of that used upon entering. For example, if the room was entered using a left turn, use a right turn when leaving the room.

This technique will help rescuers search a building systematically and give them a quick exit if they have found a victim and want to return to their starting point.

- Enter and exit in the same direction if you are to continue the search. For example, enter right, exit right.

- Enter and exit in opposite directions if you are to abort the search, or if a victim is located and is to be removed from the building. For example, enter right, exit left.

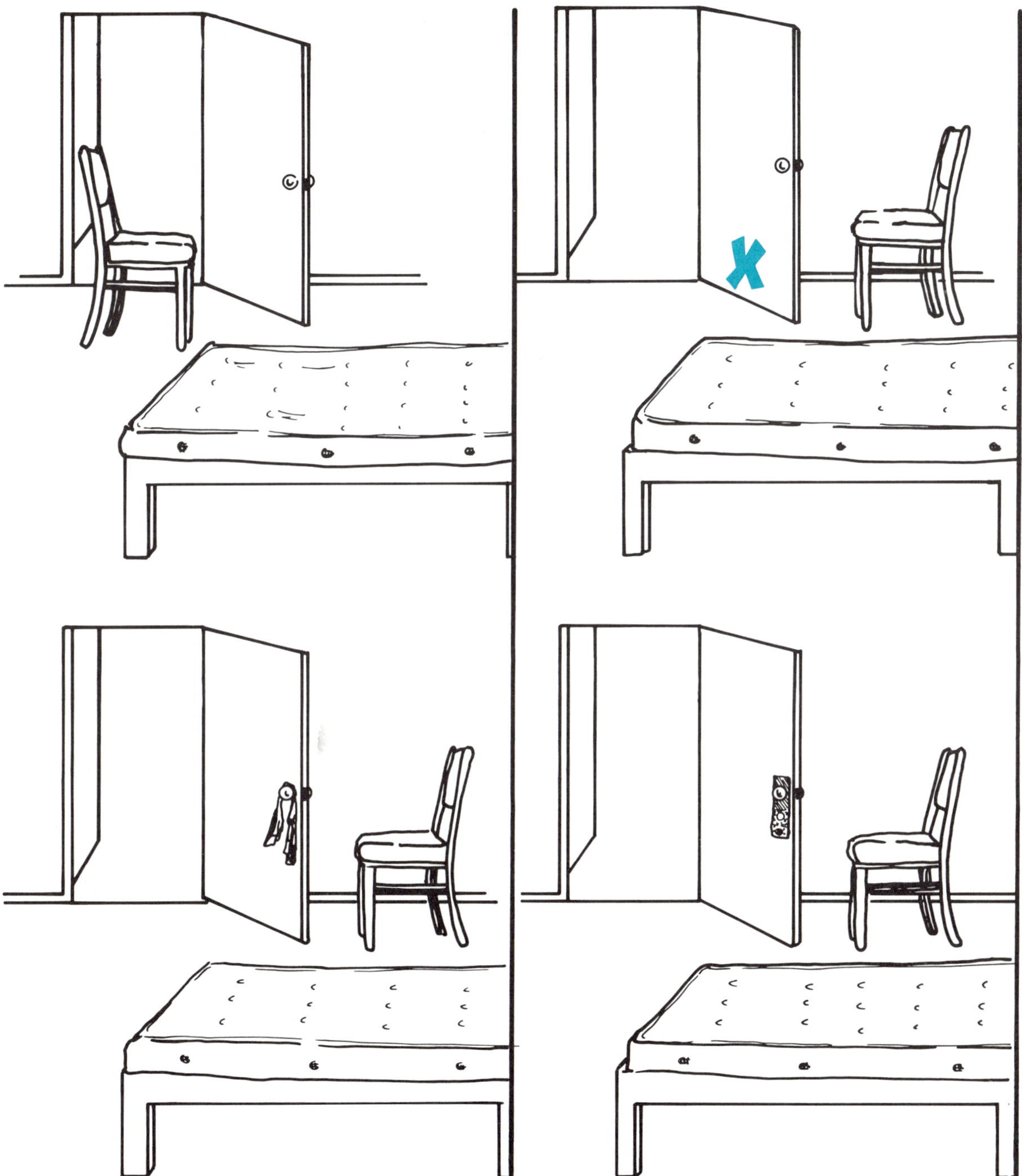

Figure 5.2 Some standard method should be used to indicate that a room and certain areas in the room have been searched. This will prevent time-wasting duplication of effort, close door after search.

If for any reason the search is aborted, this information should be reported immediately to the officer in charge. During a search for victims, negative information is just as important as positive information to coordinate a complete search.

If rescuers become trapped in the building while making a search or have located a victim, they can go to the window and call for assistance and get the victim to fresh air. Doors to rooms not involved in fire should be closed to delay the spread of fire into these rooms. Exit paths from fire buildings should be kept as clear as possible of unused hoselines, ladders, and apparatus, making the removal of occupants easier and preventing falls and possible injuries.

ROOM SEARCH

If visibility is poor, crawl on your knees or stomach to search for victims. Visibility should be clearest closer to the floor. Heat and smoke tend to rise. Crawling will also let you locate holes, obstructions, and other hazards (Figure 5.3). Follow the walls so you will not become disoriented. Windows may be opened for ventilation, but use good judgment or the fire could be extended. Feel for hot spots on the floors and walls and look for fire extensions. If you should discover fire in a room, close the door to shut off its spread.

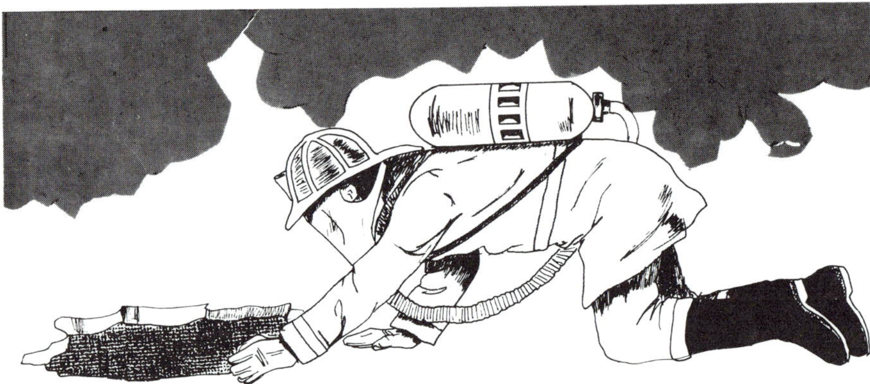

Figure 5.3 Rescuers should keep low for better vision and should feel ahead for obstacles or pitfalls.

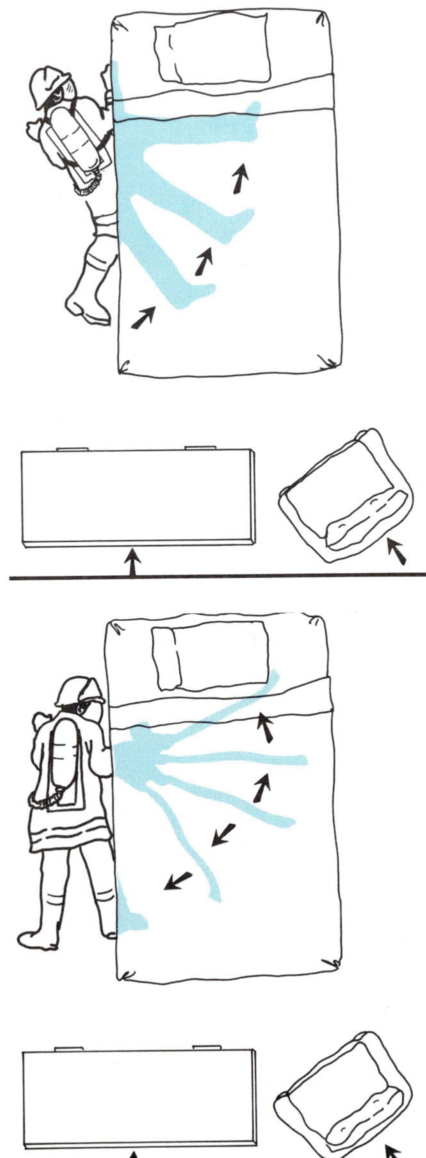

Figure 5.4 Children often hide during a fire. Sweep entire area under a bed gently with a leg or tool. Also, check behind all furniture.

Rescuers should stay in constant communication with each other, preferably by physical contact. Keep talking to your buddy to keep track of each other's condition. Pause occasionally to listen for coughing, moaning, or any sound that would indicate the location of a victim. There is also a chance you can locate the fire by hearing its crackling.

Enter the room and follow the wall around the room. Use rubber straps or other devices to keep doors from locking (see "Latch Straps," page 91). Keep in constant and direct contact with the wall, stretching out your hand or foot to cover a greater area. This will bring you back to your starting point. Then search the center of room. Check beds thoroughly on top and underneath (Figure 5.4).

Thoroughly search all areas, including behind furniture and inside closets and bathrooms. Careful consideration must be given to the search pattern so that no areas are overlooked. Figure 5.5 illustrates an inadequate search of a large room by one person. Figure 5.6 illustrates an adequate search of the same room by two searchers.

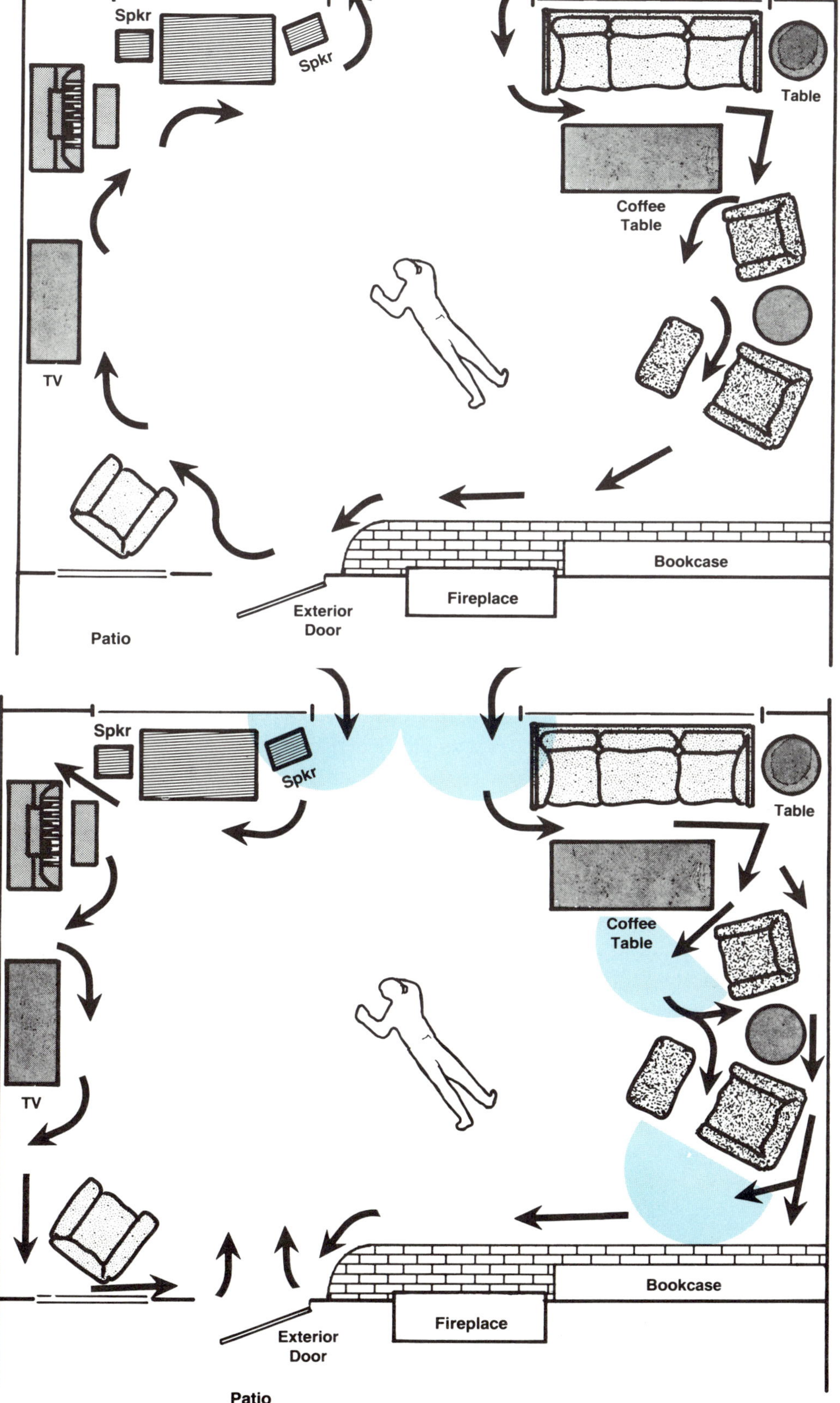

Figure 5.5 A hurried search by one person could leave a victim undiscovered.

Figure 5.6a A thorough, careful search will help find victims that otherwise would be overlooked.

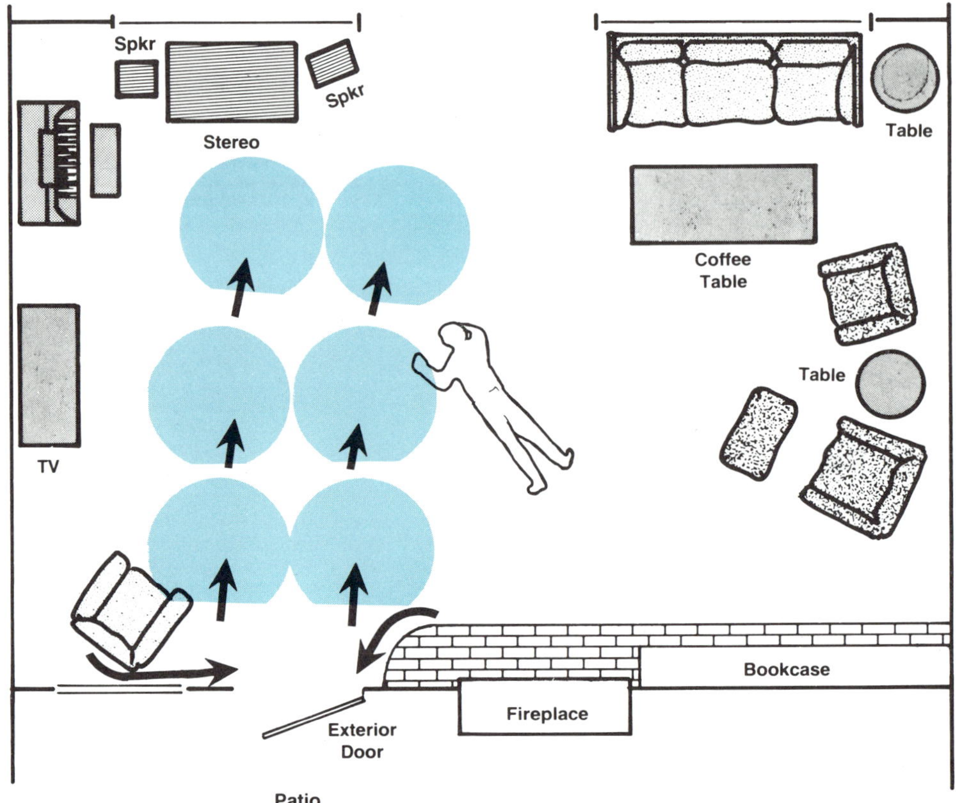

Figure 5.6b After searching the perimeter, the rescuers start into the interior of the room. Maintain visible or physical contact with an exterior wall. A rescue strap can be used between the two searchers to maximize coverage.

If smoke has reduced your visibility, use your hand, leg, or pike pole to feel for a victim under beds. You can also find victims near windows, doorways, and other escape ways (Figure 5.7). Some general items to remember in any building search are shown in Figure 5.8.

Figure 5.7 Victims may be found near windows, doors, and other exits.

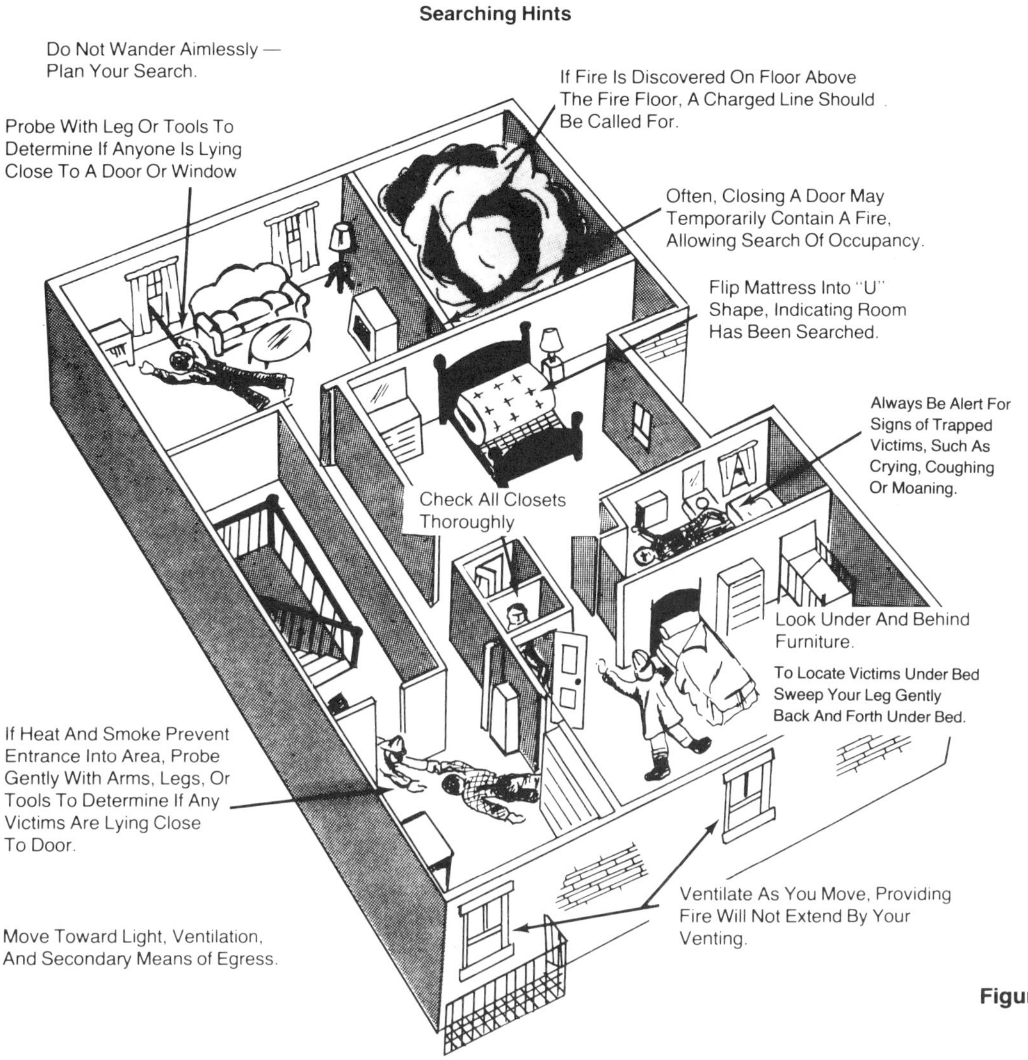

Searching Hints

Do Not Wander Aimlessly — Plan Your Search.

Probe With Leg Or Tools To Determine If Anyone Is Lying Close To A Door Or Window

If Fire Is Discovered On Floor Above The Fire Floor, A Charged Line Should Be Called For.

Often, Closing A Door May Temporarily Contain A Fire, Allowing Search Of Occupancy.

Flip Mattress Into "U" Shape, Indicating Room Has Been Searched.

Always Be Alert For Signs of Trapped Victims, Such As Crying, Coughing Or Moaning.

Check All Closets Thoroughly

Look Under And Behind Furniture.

To Locate Victims Under Bed Sweep Your Leg Gently Back And Forth Under Bed.

If Heat And Smoke Prevent Entrance Into Area, Probe Gently With Arms, Legs, Or Tools To Determine If Any Victims Are Lying Close To Door.

Ventilate As You Move, Providing Fire Will Not Extend By Your Venting.

Move Toward Light, Ventilation, And Secondary Means of Egress.

Figure 5.8 *Courtesy of WNYF.*

LATCH STRAPS

Firefighters use latch straps to keep open doors that would close and lock by themselves. The straps can be made easily and economically from a truck tire inner tube (Figure 5.9).

Figure 5.9 Latch straps are useful for keeping doors from locking and, as shown here, for indicating searched rooms.

ABOVEGROUND RESCUE

If performing a rescue in a two-story or higher building, observe the following rules. If an entrance is made through a window at ground level or by ladder at a higher elevation, be sure the floor is still intact. This can be checked by feeling with your foot or with a tool such as a pike pole. Even if the floor is still intact, check for sponginess that would indicate that it could fail at any time.

Determine floor stability

To learn the layout of fires in floors above the ground floor, study the layout of the floor below the fire. The only way to get a more accurate idea of the layout is by good pre-fire planning.

HIGH-RISE SEARCH AND RESCUE

Rescue problems in a high-rise building above the reach of fire department aerial apparatus is a challenge to all fire departments faced with the potential problem. As in other building fires, two ways to reach victims could be possible—exterior rescue and interior rescue. The high-rise problem of rescuing victims deserves much attention and thorough training for everyone concerned.

Rescue teams should assemble at the interior command post (subcommand post) two or more floors below the fire floor and check layout of floor, room numbers, exits to stairways, and distance from elevator to exits. Floors are generally laid out the same.

For example, if information obtained on arrival indicated victims trapped in specific rooms on the twelfth floor, the eleventh floor should tell you on which end of the building the rescue problem is located, what the distance is from your stairway, and about how many rooms will have to be checked. Checking room numbers on the floor below and getting some idea of the distance to be traveled will save time and energy.

Even with this information, the task will be tough. With a great amount of heat and smoke, rescue above can sometimes only be accomplished with the aid of a charged hoseline from the nearest standpipe.

OUTDOOR SEARCH

Searching techniques for persons lost outdoors may vary from case to case, depending on the area to be covered, the age and physical conditions of the lost, the time of day, and weather conditions, but the general procedure will be the same. Adhering to the same procedure increases the effectiveness of each search and the safety of the rescuers.

For a quick and successful rescue, outdoor searches must involve large numbers of people, generally from outside the fire department. Good sources of outside help are law enforcement agencies, civil preparedness groups, sheriff's posses, or similar groups. Communities with many calls per year for outdoor searches may even maintain special rescue squads for this purpose. Whatever group is used for assistance, the volunteer help must work under persons trained and competent in outdoor searches. Generally, fire department personnel should be directing the volunteer help instead of searching.

Search organization objectives

The objectives of the search organization are to determine the location of lost persons and to insure their safe return with a minimum amount of effort and time and without unduly endangering the searchers. In addition, the secondary objectives of preventing property damage, ruining crops, or frightening livestock must be kept in mind. A well-organized search program will insure that these objectives are easily met.

Since a search may be required in various locations, plans and operating procedures should be developed to establish a command post immediately. Include any portable tools that may be needed. It is best to list all the supplies needed for the worst possible search situation in the jurisdiction. Secretarial supplies and basic search equipment can usually be stockpiled in advance. Incidental hardware requiring a small investment can also be stockpiled in advance. (For a sample listing of these materials see Appendix B.) These objects should be packed in an airtight container, perhaps a galvanized trashcan, and put in a secure place. Trashcans are about the right size and the lid can provide an airtight seal. A trashcan is durable and easily handled.

Possible command post locations should be considered in advance. Each area in a jurisdiction should be examined to see if an outdoor search might sometime be necessary. Areas around campsites, outdoor recreational areas, lakes and streams, and large parks should be studied for possible command post sites. Prerequisites for the command post location include:

- *Good Communication*—Radio communication is good, but adequate equipment is often not available. Locating a command post near a pay phone is sometimes an excellent solution, especially if the location is preplanned, because the phone number of the command post can be given to involved agencies. The telephone company can have public booths prewired for command post use.

- *Adequate Shelter*—The only effective way to insure efficient command post operations is to provide shelter from the weather. The search should be planned as if it were to occur during the worst possible conditions. The stockpiled equipment listed in Appendix B includes a five-person tent, but seek the best shelter available and use the tent as a last resort. Public buildings, barns, private dwellings, contractors' trailers, or other outbuildings should be examined and agreements made for their use in an emergency. Trucks or vans have also been used effectively for command posts.

- *Adequate Parking*—Almost without exception, searchers will have to travel some distance to the command post. A parking area for vehicles must be considered. Parking on the side of roadways is not acceptable because of the danger from moving traffic.
- *Reasonable Proximity* — It is important that the command post be close to the search area for logistical reasons. However, command posts have been effectively managed several miles away from the search area when other factors have taken precedence.

COMMAND POST DUTIES

Coordination needed for methodical search

Coordinating an open-area search is difficult. Each section of the area has to be methodically searched. Plans must be made so no place is overlooked and efforts are not duplicated. In addition, the welfare of the searchers must be kept foremost in mind. These basic objectives of command are handled through three areas of responsibility: operations, communications, and logistics.

OPERATIONS

The entire search is coordinated by the operations officer, usually a chief officer who heads the section that assigns personnel to search areas and monitors the search progress. This is the line section of the search organization. Specific duties include:

- *Gathering Information*—Information is the foundation of good decisions. Information about the lost person must be gathered and distributed. This information should include a description of the person's physique, type and color of clothing, health, personal habits, last known location, and anything else that might assist searchers. Information about the weather must be constantly monitored and updated.

Everyone must be briefed

- *Briefing Searchers*—The search units must be briefed about their specific function. They must be advised of their authority and responsibility. They must know how long they have before returning to the command post to report. They must understand that if they fail to return at this time, people will be immediately dispatched to their area with the assumption that they are in trouble. Of course, they must also be told all the information about the lost person.

- *Controlling Personnel*—The operations officer must require each searcher to register before being assigned to a search unit. Each must also be checked out when the search ends. The operations officer must insure that no one participates in the search without proper clothing. The operations officer should instruct search unit leaders that persons cannot participate unless they are assigned by the operations section.
- *Coordinating Auxiliary Operations*—The operations officer coordinates with the communications and logistics sections to meet the needs of the operation.

COMMUNICATIONS

The communications section handles all messages and prepares public information releases. This may only involve answering or dialing the phone, or it may involve sending a handwritten message several miles into the field. The communications people should make every effort to expedite messages. If radio equipment is available, the communications section should establish a base station and control the radio traffic. If radios are to be used, a frequency must be selected that will not interfere with other emergency traffic. Administrative or mutual-aid frequencies are often used to leave the fireground frequency open.

News bulletins should be periodically prepared, if possible, before journalists ask for them. Requests for information may come when preparing a release is difficult. News releases put together hurriedly may lack specifics and can be misunderstood by the journalists. If a good working relationship already exists between journalists and the local fire department, this task will be less difficult. A good release should contain as many facts as possible and the name of a person who can be quoted. Spokespersons should never speculate or draw conclusions, but should be willing to clarify and add to the information in the release if it can be done factually. Above all, they should not make the situation seem better than it is.

Prepare news bulletins periodically

Always try to cooperate with the journalists and understand that reporters and photographers have a right to be at the scene of any newsworthy event. They are performing a public function, but they may not realize the dangers into which they put themselves and others by wandering around in unsafe areas. The communications section must control the situation and help protect these observers.

Cooperate with journalists

There may be situations where help from the press is vital. For example, it may be necessary to notify the public immediately or it may be best not to alert the general public right away. If a good working relationship exists, reporters can generally be depended on to cooperate.

LOGISTICS

The logistics section provides material for the operations. The logistics section should be familiar with the resource book and begin contacting sources of necessary equipment. However, no procurement should be made until requested by the operations officer. Sometimes equipment can be located and staged in the command post area for immediate deployment. Items of immediate consideration include heating equipment in cold weather, lighting equipment, food, and communications equipment.

The logistics section is also charged with providing transportation, arranging to move search units to the search location from the command center or from location to location, and arranging to move supplies from their source to the location of use.

The logistics section must be aware of their requisition authority and financial ability. Often, special accounts are set up with local suppliers to provide credit for such emergencies. Accounts should be set up with such businesses as food and fuel suppliers or equipment rentals.

Principle of Open Field Search

Open field search is the methodical and complete visual inspection of open terrain for a lost or injured person, or for signs, indications, and marks of the person's movement. Large areas are divided into smaller sections along natural or man-made boundaries such as creeks, ridges, roads, and fences. These areas are methodically searched in sweeps by lines of search units spaced abreast. Each search unit is assigned a certain section to check. The areas around the last known location are generally checked first, then the areas away from that location.

Search kits for units

A search unit is similar to a company, but generally only one or two persons in a unit will be from the fire department. A unit can involve five to ten persons. The unit leader is usually a fire department member. Each search unit should have a search kit:

- A pad and pencil
- Two rolls of vinyl tape
- A whistle
- A compass
- A map of the search area with the unit's area marked and their return time indicated
- All the available information possible about the lost person

Units are directed by unit leaders, who are responsible for the unit. The leader's main responsibilities are to insure the welfare of his/her personnel, to make sure the search is methodical and complete, since failure to check one small area may make the search fail, and to protect the property and livestock of others from damage that may result from the search unit's actions. Some guidelines to follow:

Search guidelines

- Check cultivated fields from the perimeter. By walking around the outside, searchers can find footprints leading into or leaving the field. If no footprints can be found leaving the field, a select group can check the field while being careful not to damage the crop.
- Go under fences when possible. Climbing over usually pulls the wires loose. Use gates if accessible, but make sure they are left as they were found.

THE SEARCH PROCEDURE

The search unit leader should assemble the unit at a corner of the search area assigned to that unit, and the unit should then be dispersed abreast along either border extending from that corner. The spacing between unit members will be determined by the physical characteristics of the terrain. In thick underbrush the spacing may be arm's length. In large open fields the spacing may be up to 100 feet (30 m). The only criterion is that the area between unit searchers be completely visible. The unit leader takes one end of the unit and appoints an assistant leader to take the other end.

When the unit is in position, the leader signals the squad to move forward by raising a hand straight above and dropping it forward. The squad then moves in a line across the search area (Figure 5.10). Leaders must search their areas and watch their unit personnel to make sure they stay in line. If the line spreads out, areas can be missed, if this happens, the search leader should blow the whistle provided in the search kit, and the unit should align itself between the unit leader and assistant leader and wait for the leader's signal to move forward.

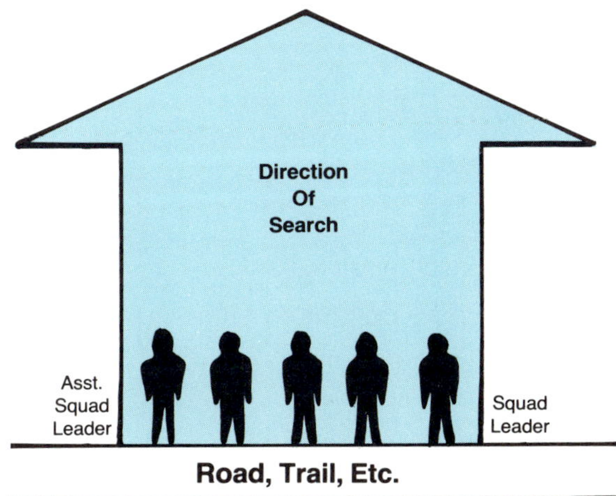

Figure 5.10 Deployment of personnel for first sweep.

The search leader and the assistant search leader must see that each thicket or bush is searched carefully. Lost persons tend to seek the shelter of these areas. Usually they hide in or under something, not necessarily to avoid being found but from their natural instinct to hide from the unknown—the night, the cold, or fear itself.

When the unit has reached the far boundary of the section, the leader blows the whistle to align the unit. When searchers have come into position, the leader faces them and signals for them to move single file along the edge of the search section by holding both hands above the head while facing the squad and dropping the arms forward to shoulder height. The unit should then file along the termination line until the unit leader is in the spot vacated by the assistant leader (Figure 5.11). When the unit is in the new position, the unit leader signals for forward movement and the unit sweeps back toward the original section boundary. The sweeps are continued until the assigned area has been covered (Figure 5.12).

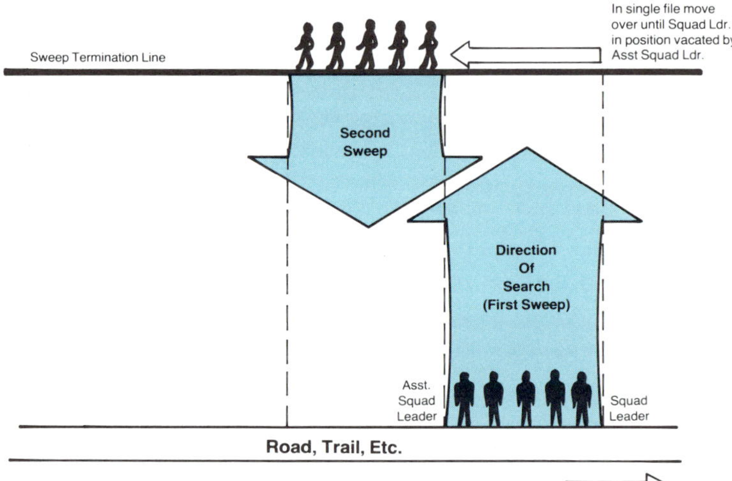

Figure 5.11 Method of search unit moving over for a second sweep.

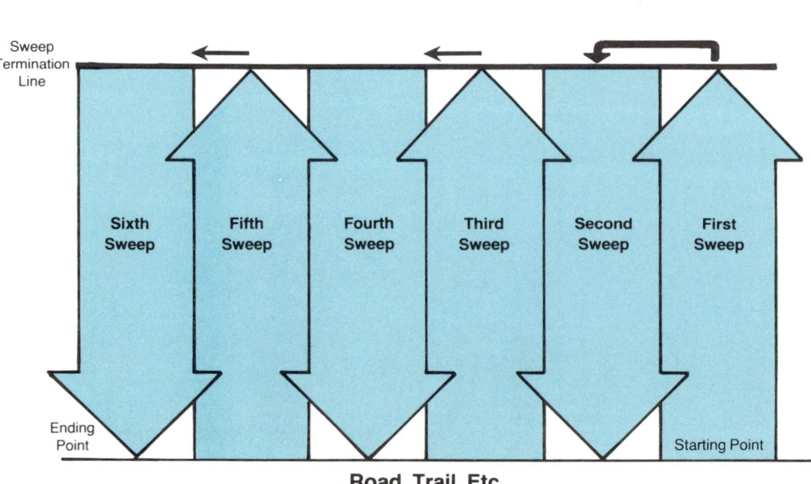

Figure 5.12 Movement of search unit over search area.

MARKING THE SEARCH AREA

Marking the search area is necessary. Failure to keep track of searched areas has proven catastrophic in some instances. Two rolls of marker tape are included in each search unit kit. These tapes can be used to mark a section of the area so its status can be easily determined by other persons. This is important so an area is not searched twice, wasting effort or time. It is even more important to insure all areas are searched.

Figures 5.13-5.18 (on next two pages) show methods of marking. These marks are placed by the unit leader and assistant unit leader (each should have one roll of the vinyl tape). When the unit is in position to begin the search, the unit leader places two hatches on the ground and the assistant places one hatch on the ground. As the unit moves across the field, the unit leader and assistant unit leader leave marks at intervals suitable to the terrain to mark the sweep path.

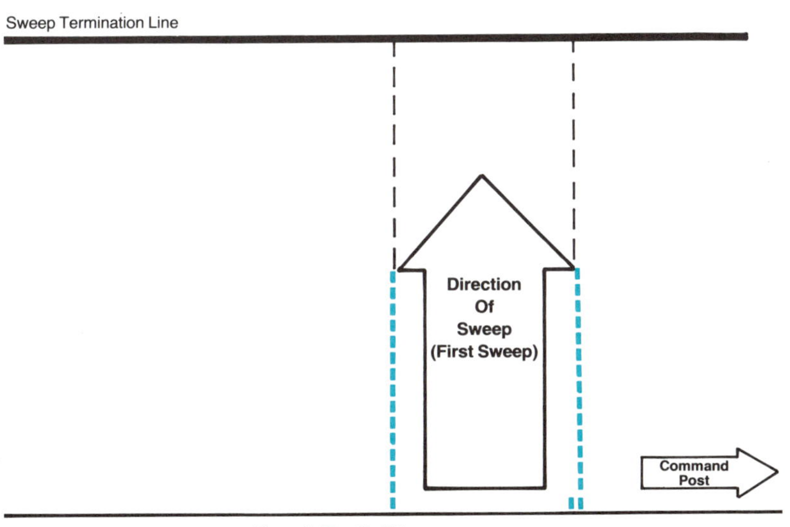

Figure 5.13 Placement of markers by leader and assistant leader during first sweep of open area.

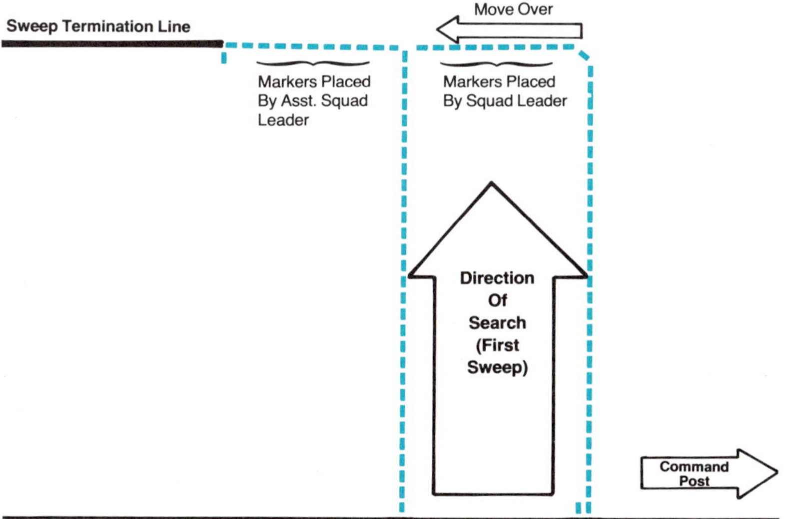

Figure 5.14 Placement of markers for second sweep of open area.

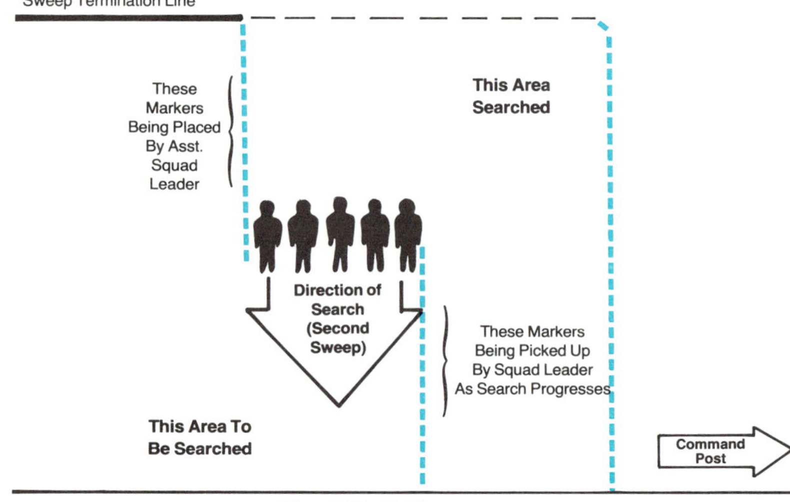

Figure 5.15 Placement and pickup of markers during second and subsequent sweeps of open area.

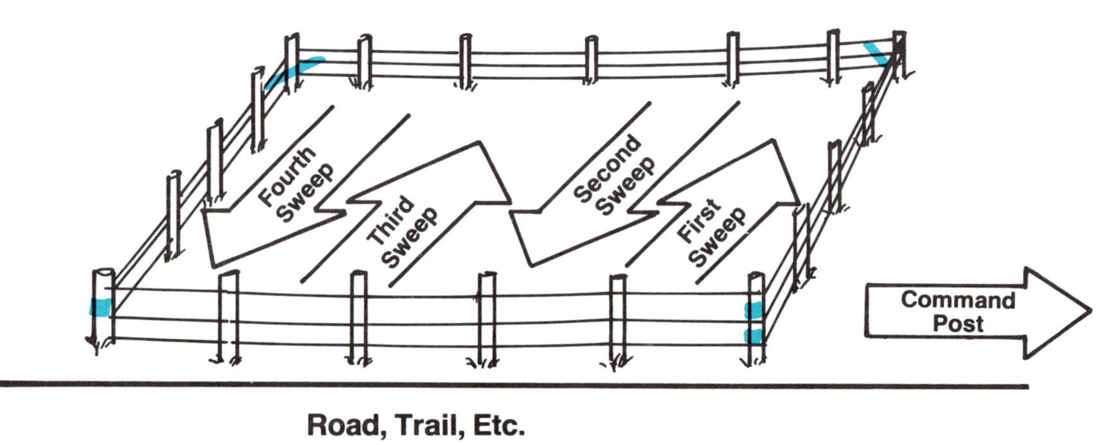

Figure 5.16 How to mark a fenced field.

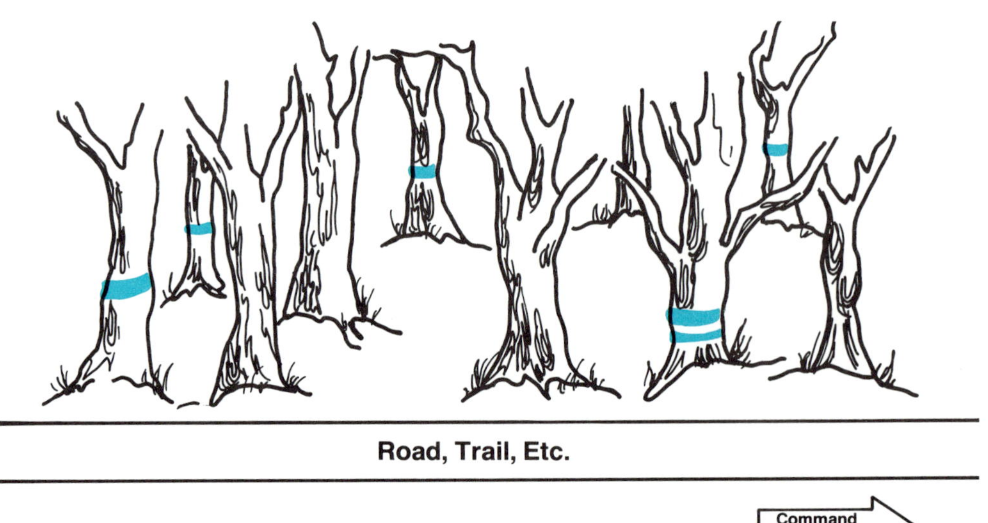

Figure 5.17 Marking a wooded area is the same as for any other except the tapes are fastened to trees.

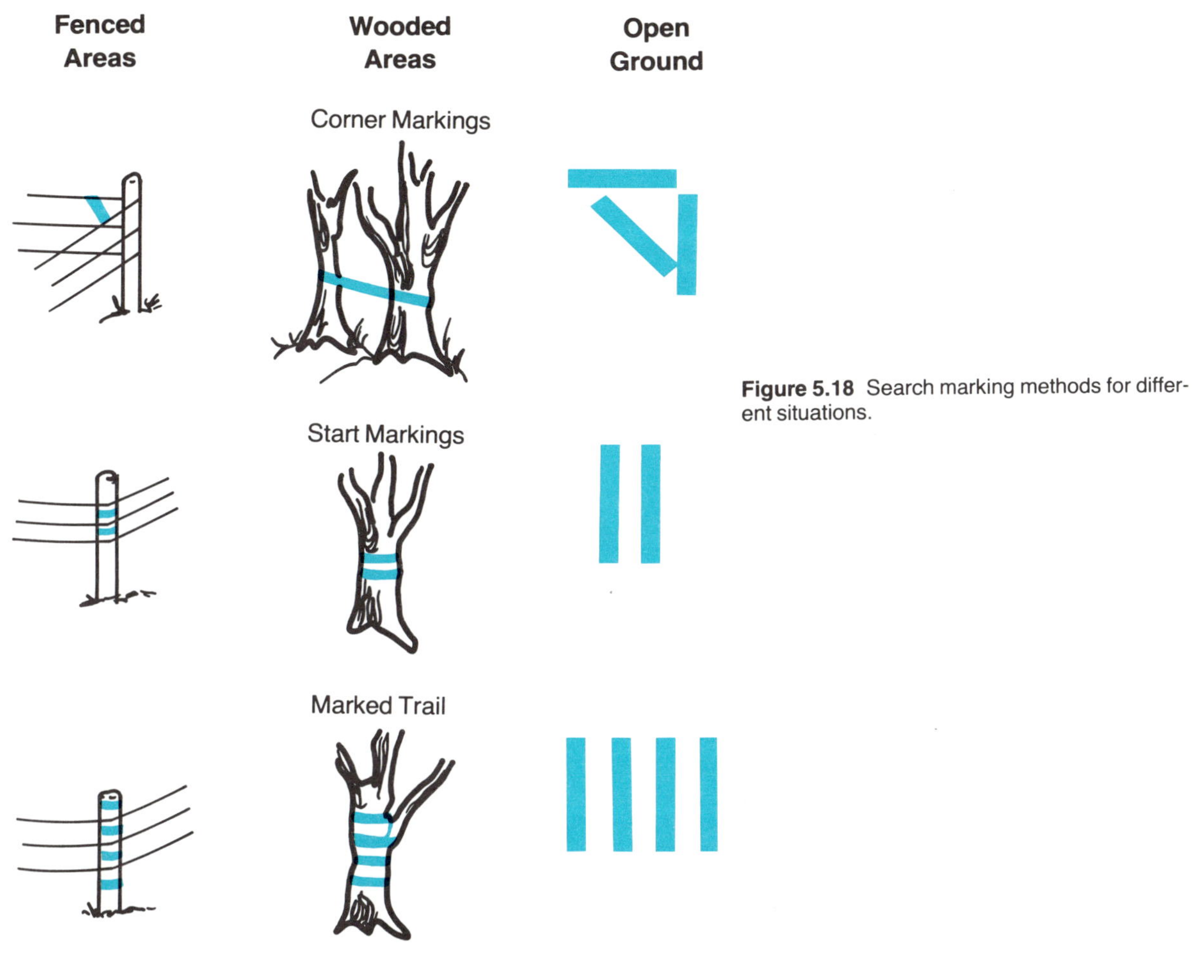

Figure 5.18 Search marking methods for different situations.

At the termination line, the unit leader marks the edge with an edge marker. The assistant unit leader leaves one hatch. When the sweep is reversed, the unit leader places an additional hatch on the assistant's single hatch to show the corner or side of the search area. Major trails should also be marked for future reference. *The above material on Open Field Search is used with the permission of the copyright holder, Stoyan Russell.*

Dogs Used in Search

In many areas of the country, there are people who have trained tracking dogs and who are willing to help if contacted. The most successful tracking dog is the bloodhound; however, the German shepherd has also been used extensively on fresh trails. A good tracking dog can save hundreds of hours of work if used properly.

MOVING THE VICTIM TO SAFETY

Helping a Victim Walk

A semiambulatory victim may only require help to walk to safety, probably the least laborious of all transportation methods. One or two rescuers may be needed, depending on how much help is available and the size and condition of the victim.

ONE-PERSON METHOD

If victims are prone, the rescuers should assist them to a sitting position and then to their feet. Grasp one arm, place it over your shoulder, and secure it by holding the wrist. Place your other arm around the victim's waist and help the victim walk (Figure 5.19).

TWO-PERSON METHOD

The two-person method takes the same positioning, but two rescuers perform the task instead of one. Two rescuers can support the victim's entire weight (Figure 5.20).

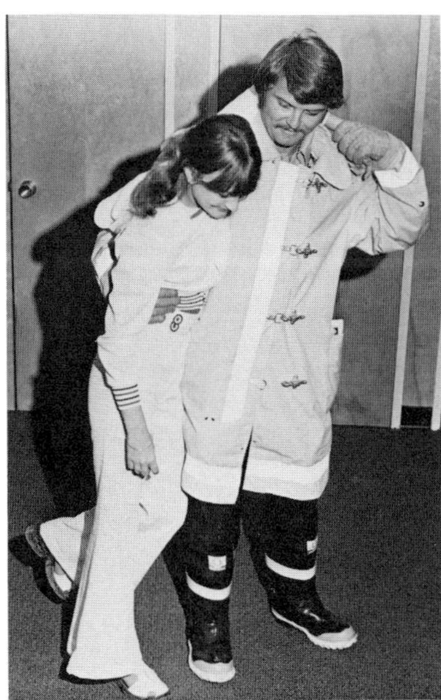

Figure 5.19 One person assisting victim to walk.

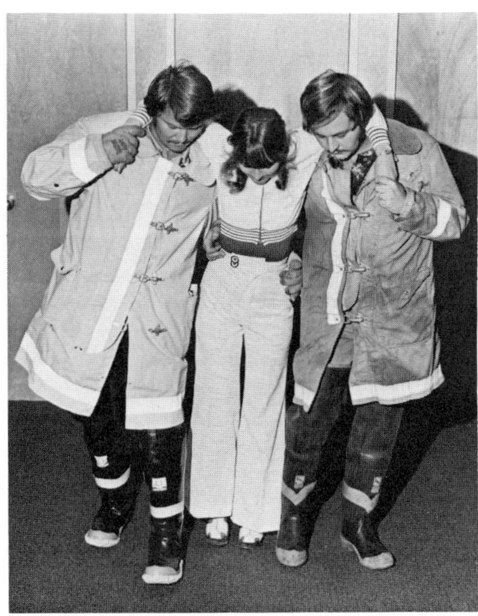

Figure 5.20 Two persons assisting victim to walk.

Carrying a Victim

When unconscious or unable to walk from injury or illness, the victim must be carried or dragged out of the dangerous environment. A one-person carry is not always practical, especially if the victim is unconscious or extremely heavy. Two- and three-person carries are easiest and should be used.

SEAT CARRY

The seat carry is a two-person method of transportation that can be used for a conscious or unconscious victim. Follow the steps below.

Step 1: The rescuers kneel on either side of the victim near the hips. They raise the victim to a sitting position, and steady the victim with their arms around the back. The rescuers should grasp each other's shoulder or upper arm to secure the back support. (Figure 5.21).

Step 2: The rescuers should place their other arms under the victim's thighs and clasp wrists (Figure 5.22). Adjust the upper arms, if necessary, to insure a comfortable and secure back support.

Step 3: Slowly rise together and transport the victim in this position (Figure 5.23).

Figure 5.21 To begin the seat carry, get victim to sitting position and link arms firmly behind victim's back.

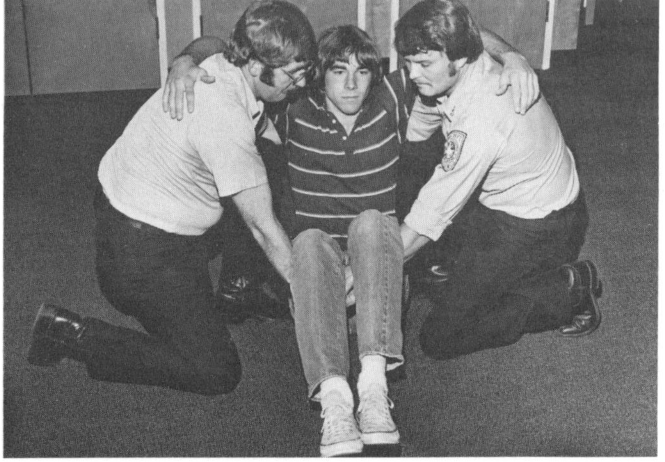

Figure 5.22 Link free arms underneath victim's thighs.

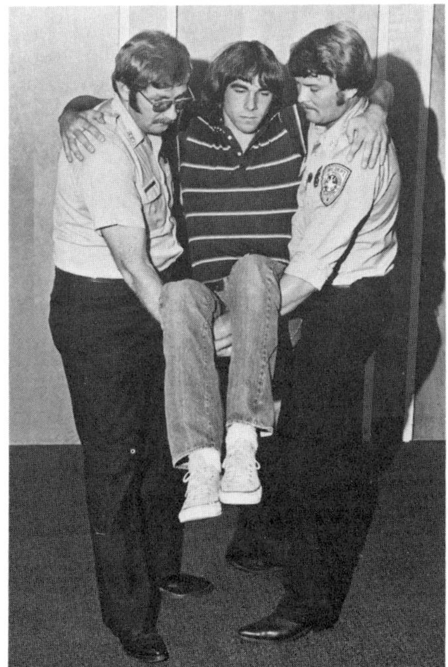

Figure 5.23 Lift and carry the victim to safety.

EXTREMITIES CARRY

The extremities carry is another two-person carry that is easy to do with conscious and unconscious victims. The steps are listed below.

Step 1: One rescuer stands at the head of the victim, and the second rescuer stands at the feet.

Step 2: The rescuer at the head kneels and slips the arms under the victim's arms and around the chest, grasping the victim's wrists (Figure 5.24).

Step 3: The rescuer at the feet kneels with feet together between the victim's legs. This rescuer grasps the victim under or just above the knees (Figure 5.25).

Step 4: The two rescuers then stand and carry the victim to a place of safety (Figure 5.26). (Remember to use your leg muscles when coming erect.)

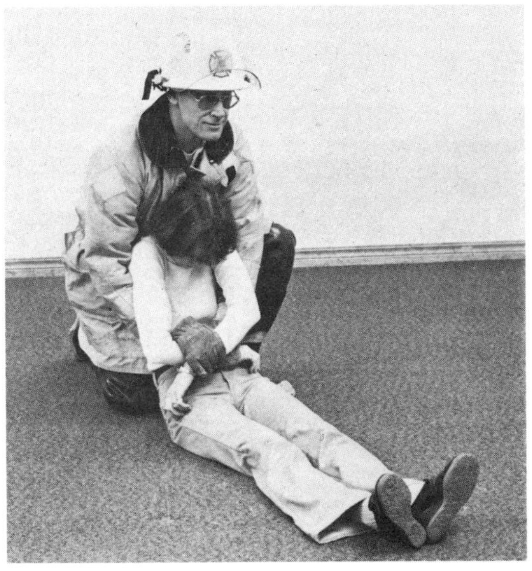

Figure 5.24 To begin the extremities carry, get victim to sitting position and grasp victim's wrists from behind.

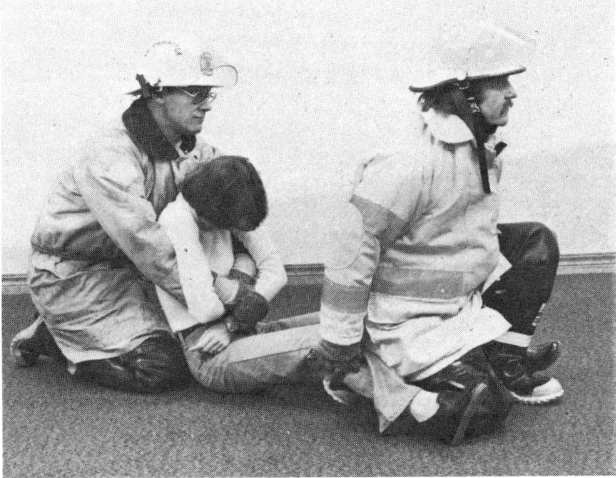

Figure 5.25 The second rescuer kneels between victim's legs and grasps victim under the knees.

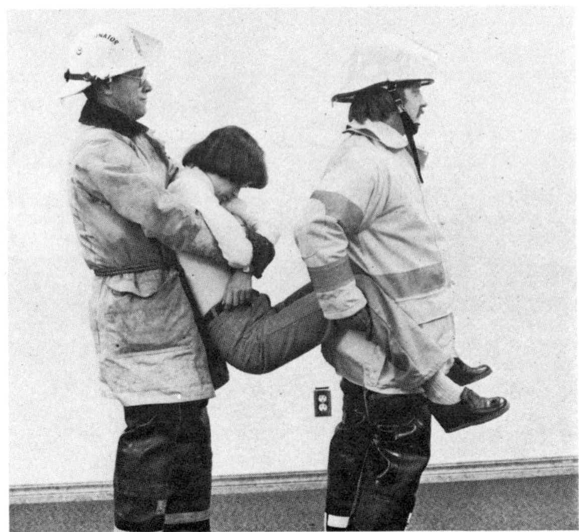

Figure 5.26 The victim is then lifted and carried to safety.

CRADLE-IN-ARMS CARRY

This carry is suitable for a child or small adult, but is difficult for carrying an average or large unconscious adult. Simply pick up the victim in your arms. Kneel and support the victim with one arm under the thighs and the other arm under the upper back. Roll the victim into the hollow of your arms and use the leg muscles to stand up (Figure 5.27).

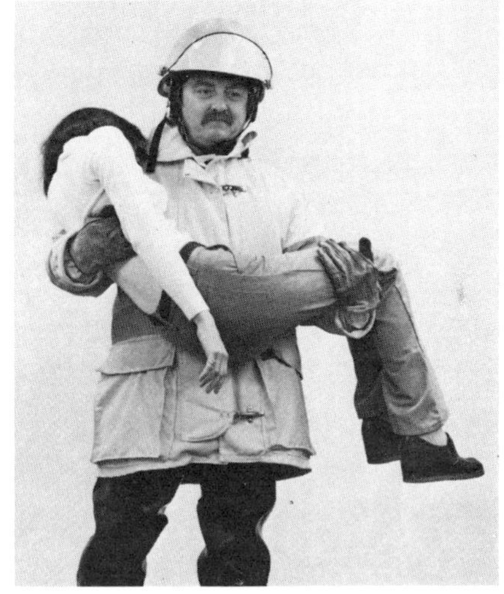

Figure 5.27 The cradle-in-arms carry is useful for children and lightweight adults.

CHAIR CARRY

The two-person chair carry consists of the victim being seated in a sturdy chair with one rescuer in front of the chair and one at the rear. The rescuer to the rear puts the hands over the victim's shoulders and through the armpits to grasp the back of the chair (Figure 5.28).

The chair can then be slightly tilted backward on its rear legs while the rescuer in front, back to the victim, squats, and grasps the front legs of the chair with the victim's legs between the rescuer's arms (Figures 5.29 and 5.30). The rescuers should be sure that both are ready to lift the chair with the victim. Then the rescuer at the rear gives the command: "Ready, lift."

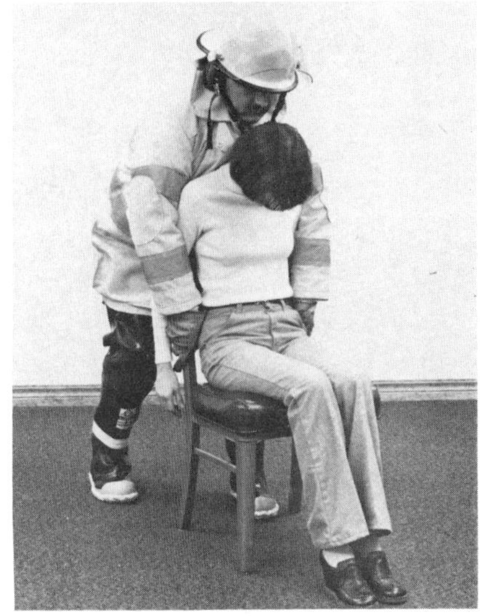

Figure 5.28 Begin the chair carry by reaching over the victim's shoulders and grasping the back of the chair.

The victim should be reassured and encouraged to grasp the frame of the chair.

If the victim is unconscious, follow the steps below.

Step 1: Turn the victim on the back.

Step 2: One rescuer, between the legs, grasps the knees and raises the victim until the legs, buttocks, and back are high enough for the second rescuer to slip the chair under the victim.

Step 3: Raise the back of the chair to a 45-degree angle, and pick up the chair with one rescuer carrying the back and the other carrying the legs.

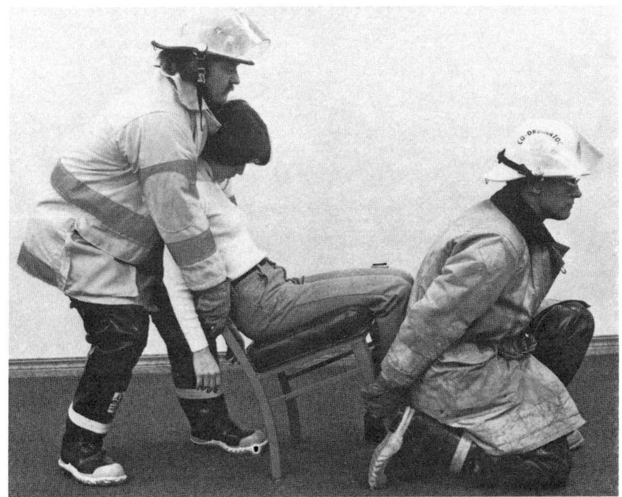

Figure 5.29 The first rescuer tilts the chair backward and the second rescuer grasps the bottoms of the front legs.

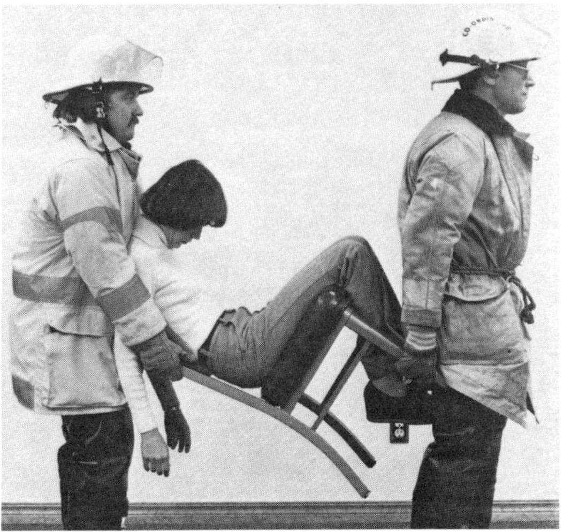

Figure 5.30 The chair and victim are then lifted and carried to safety.

If the chair cannot be slid under the victim, turn the victim on the back. Each rescuer then firmly grasps the victim under the arms and gently lifts or swings the victim onto the chair. Follow Step 3 on the preceding page to move the victim to safety.

An alternative two-person chair carry that may be used consists of the victim being seated in a sturdy chair with one rescuer on each side of the chair. Both rescuers bend and grasp the seat of the chair near the front, and the back of the chair (Figure 5.31). Both rescuers lift the chair and victim in unison (Figure 5.32). If the victim is unconscious, the chair may be tilted backwards sufficiently to prevent the victim from falling out.

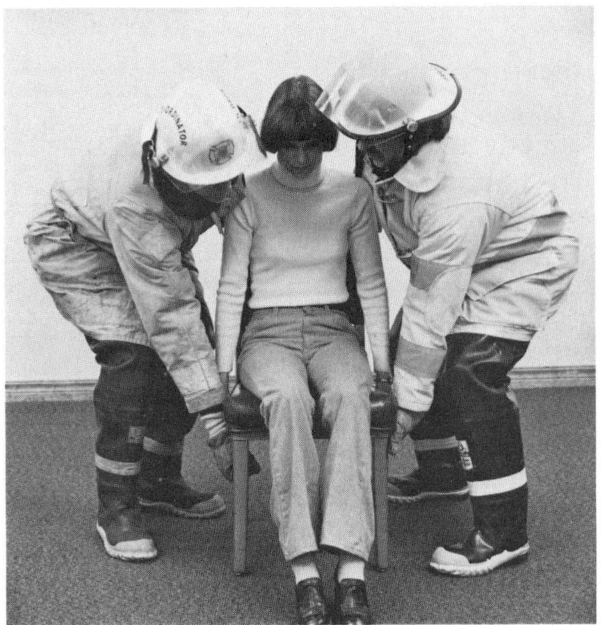

Figure 5.31 Alternative chair carry: Rescuers position themselves on opposite sides of chair, grasping front of chair seat and chair back.

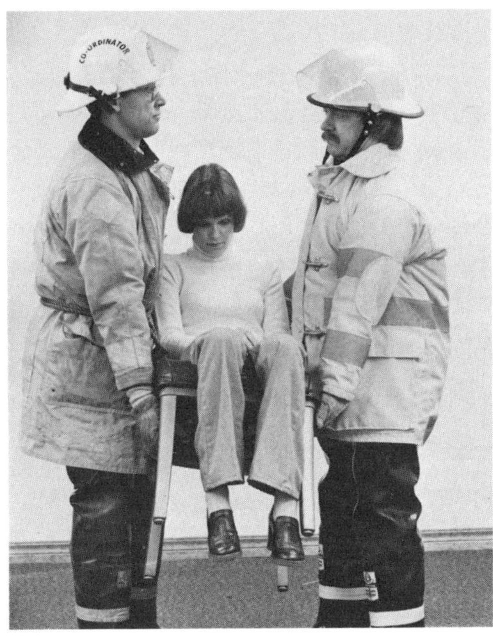

Figure 5.32 The chair and victim are then lifted and carried to safety.

THREE-PERSON CARRY

Even though this is called a three-person carry, a fourth person is needed to help load the victim. This carry is used for a person with serious injuries when a stretcher is not available. The steps for executing this carry are listed below.

Step 1: Three persons are positioned on one side of the victim—one at the head, one at the waist, and one at the feet.

Step 2: All three should kneel on the knee closest to the victim's feet.

Step 3: The person at the head slips one hand under the neck and one about the middle of the chest. The second person places one hand under the waist and one just below the hips. The third person puts the arms completely under the legs at the knees and ankles. The fourth person, positioned on the other side of the victim, puts the hands deep under chest and hips to help raise and balance the victim.

Step 4: On signal, all three raise the victim with the help from the fourth (Figure 5.33) and place the victim on their knees (Figure 5.34).

Step 5: The three rescuers then stand and roll the victim up against their chests (Figure 5.35). This will keep the victim from suffering further injuries while being carried to safety.

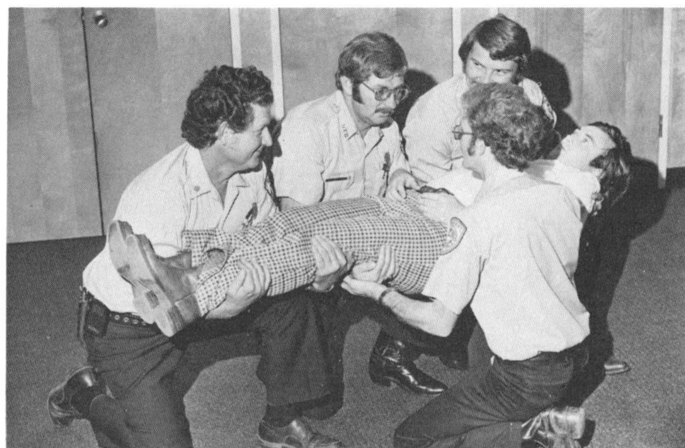

Figure 5.33 On signal, all three raise the victim with help from the fourth person.

Figure 5.34 The victim is supported on the rescuers' knees as they prepare to lift.

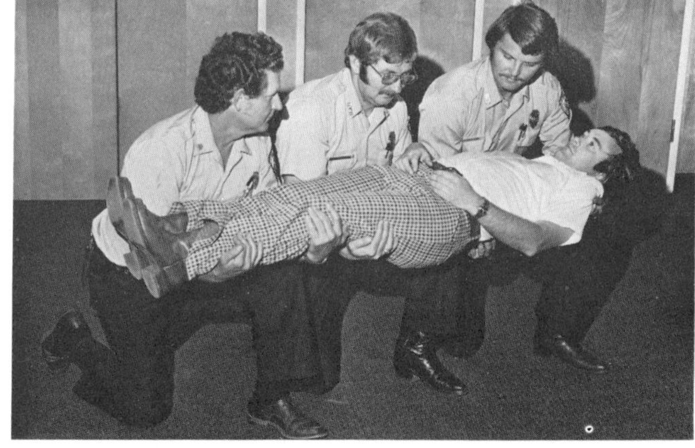

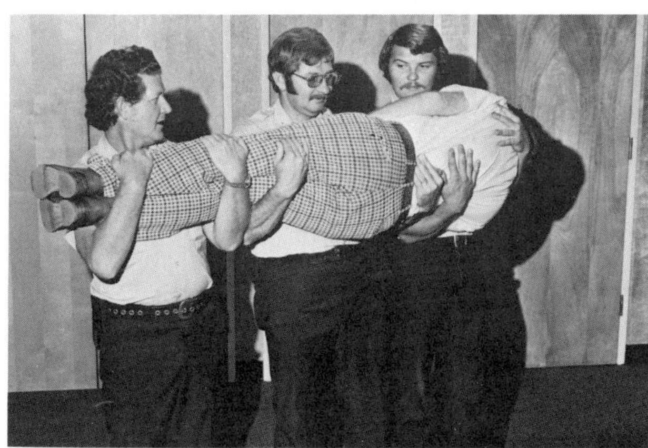

Figure 5.35 Rescuers then stand and roll the victim up against their chests.

LONE RESCUER LIFT AND CARRY

Sometimes a lone rescuer will have to carry a victim but will have difficulty getting the victim up. The following method, injuries permitting, is often successful in such a case.

Step 1: Push the victim's feet close to the buttocks and hold them in place with a foot (Figure 5.36).

Step 2: Grasp the victim's hands and rock the victim up and down several times, the distance of the rocking becoming greater each time, for momentum (Figure 5.37).

Step 3: When ready, at the top of an upswing the rescuer jerks the victim up and onto a shoulder (Figure 5.38).

Figure 5.36 Push the victim's feet close to the buttocks.

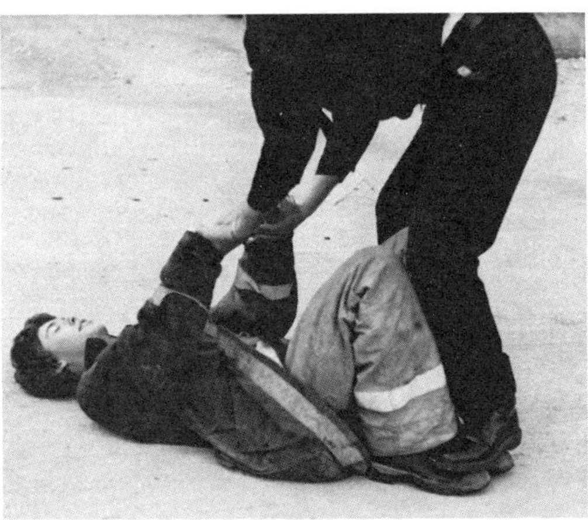

Figure 5.37 Brace the victim's legs with one foot and rock the victim up and down several times for momentum.

Figure 5.38 At the height of an upswing, the rescue jerks the victim up onto the rescuer's shoulder and carries as shown here.

Dragging a Victim

Carries are usually preferred to drags because they support the victim more and lessen the chance of further injury. Dragging a victim, however, may prove more effective and practical when only one rescuer is available and when speed is the prime concern. The following drag is relatively easy for one person to execute, regardless of the victim's size or state of consciousness.

BUNKER COAT OR BLANKET DRAG

This drag can be executed by placing a blanket, bunker coat, or similar object under the victim. Follow the steps below.

Step 1: Place a bunker coat or blanket beside the face-up victim (Figure 5.39) and gather one edge close to the victim's side.

Step 2: Roll the victim toward you and, while supporting the victim, gather the coat or blanket underneath (Figure 5.40). Roll the victim onto the coat or blanket and straighten it out (Figure 5.41).

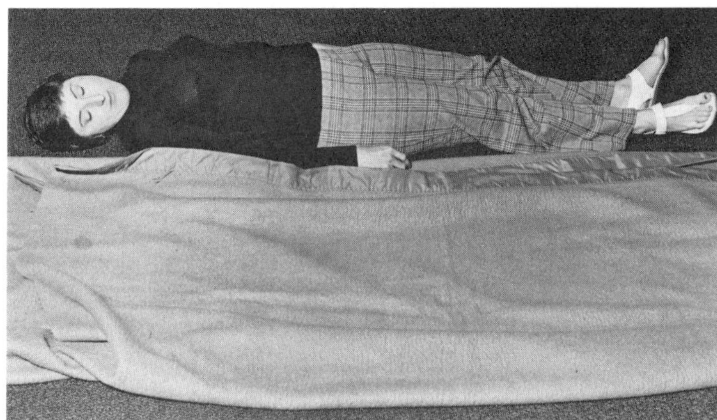

Figure 5.39 Position blanket or bunker coat next to victim.

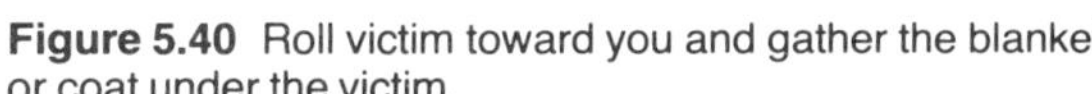

Figure 5.40 Roll victim toward you and gather the blanket or coat under the victim.

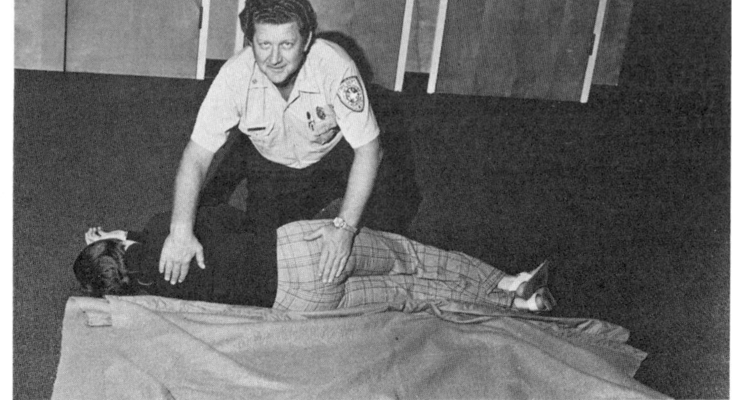

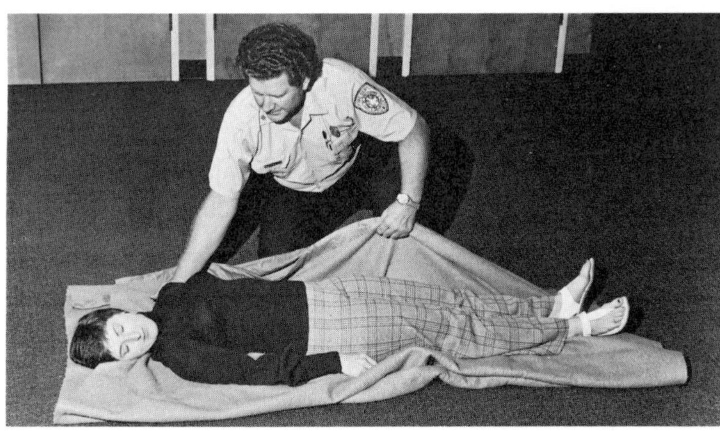

Figure 5.41 Roll victim back onto blanket or coat and straighten it out.

Step 3: Grasp the coat or blanket on each side of the victim's head and raise enough to clear head and shoulders off the floor. In this manner drag the victim to safety (Figures 5.42 and 5.43).

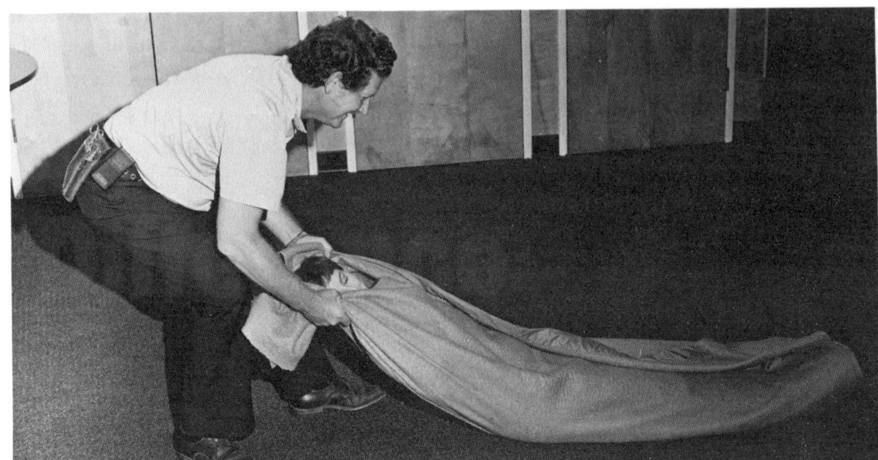

Figure 5.42 Grasp blanket or coat, lifting victim's head and shoulders off the floor, then pull victim to safety.

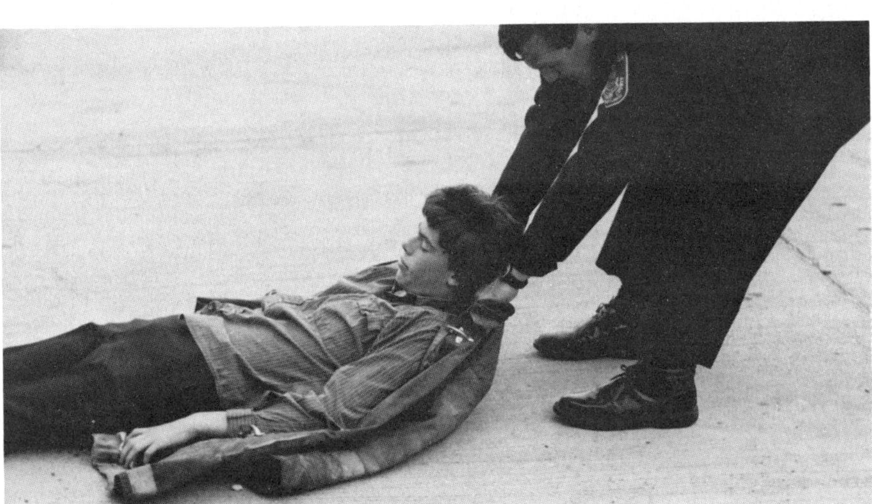

Figure 5.43 When using a bunker coat, grasp the collar yoke or shoulder area of the coat.

Rescue In Situations Involving Elevation Difference

Chapter 6

NFPA STANDARD 1001
RESCUE
FIRE FIGHTER I

3-12.3 The fire fighter shall define the uses of a life belt.

RESCUE
FIRE FIGHTER II

4-12.3 The fire fighter shall demonstrate the techniques of preparing a victim for emergency transportation by using standard available equipment, or by improvising a method.

Reprinted by permission from NFPA Standard No. 1001, *Standard for Fire Fighter Professional Qualifications*. Copyright © 1981, National Fire Protection Association, Boston, MA.

Chapter 6

Rescue in Situations Involving Elevation Difference

Rescue typically brings to the mind's eye many pictures: searching for the victim, freeing a trapped victim, carrying the unconscious victim, and finally moving the victim to safety. This section addresses special considerations for moving a victim from one elevation to another. Occasionally, the firefighter may be fortunate enough to have specialized equipment such as aerial ladders or elevating platforms to make the task relatively simple. More often, however, firefighters must use basic resources for rescue—ingenuity, ropes, ladders, and personnel.

Not all techniques are described below, but understanding the techniques presented will help the firefighter adapt to most situations.

USING STAIRWAYS IN RESCUE

Stairways better than ladders

Using a ladder is a good way to remove people from upper floors, but whenever possible, the stairway is still the easiest. This is another reason why firefighters should protect stairways and keep them clear for exit. There are many reasons to use the stairs when possible:

- More people can be removed or guided down a stairway than down a ladder.
- They can be removed faster.
- Some people will resist getting on a ladder at some heights.
- The old and invalids would have to be carried down ladders.
- Victims on ladders are exposed to falling objects.

- It is easier to transport an unconscious person on a stairway — you can stop and rest or change your grip.
- The possibility of falling is less.
- You may only have to direct the victim to a floor below the fire if the building is fire resistive.

ASSISTING A VICTIM DOWN A LADDER

Firefighters engaged in ladder rescues may discover victims with different injuries and states of consciousness. An approaching fire will require immediate evacuation even though additional help may not be available. Since even a conscious victim is generally unaccustomed to climbing down a ladder, great care must be exercised to prevent accidental injury while descending. If fire conditions permit, evacuate the victim by another means — down an inside stairway or even up to the roof and across a connecting roof where the stairway is safe. If it is necessary to bring a person down a ladder, follow the procedures below.

Placing Victim on Ladder

When it is known in advance the ladder will be used for a window rescue, the ladder tip is raised only to the sill (Figure 6.1). This gives the victim easier access to the ladder. All other loads and activity should be removed from the ladder, which should be securely anchored at both the top and heel if possible.

Figure 6.1 When using a ladder for rescue through a window, bring the tip only to the sill.

Use three or four firefighters on a ladder rescue. For ground ladder rescues, at least two firefighters should be in the building and one on the ladder. If the victim is conscious, the firefighters in the building should lower the victim feet first from the building to the ladder. The rescuer on the ladder supports the victim and descends the ladder. The rescuer descends first, keeping both arms around the victim, with hands in front of the victim for support in case the victim slips or passes out (Figure 6.2). Reassure the victim constantly while descending because the victim probably will be nervous and panicky.

An unconscious victim is held on the ladder in the same way as a conscious victim except that the body rests on the rescuer's supporting knee. Place the victim's feet outside the rails to prevent entanglement.

Figure 6.2 Bringing a conscious person down a ladder.

Figure 6.3 Bringing an unconscious person down a ladder.

Another way to lower an unconscious victim involves the same hold by the rescuer, except that the victim is turned around to face the rescuer. The position lessens the chance the victim's limbs will catch between the rungs (Figure 6.3).

In another method, the victim is lifted out to the rescuer on the ladder. The victim is supported at the crotch by one of the rescuer's arms and at the chest by the other arm. The rescuer descends with the aid of another firefighter (Figures 6.4 -6.6).

If the victim is heavy or several rescuers must help, two ground ladders can be placed side by side. One firefighter descends first and supports the victim. A second firefighter on the other ladder descends slightly above the victim, supporting the upper torso.

When bringing small children down, cradle them across your arms.

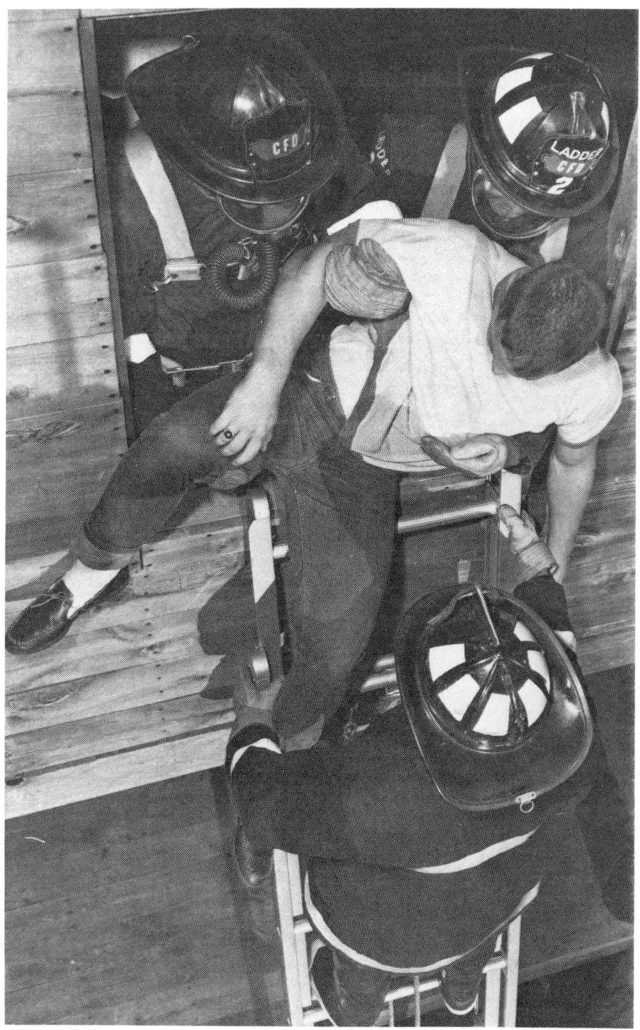

Figure 6.4 Carefully lower the unconscious person to the rescuer on the ladder.

Figure 6.5 The victim's legs are placed astraddle one of the rescuer's arms, the upper torso supported by the other arm.

Figure 6.6 With a second person backing the rescuer, the victim is brought down the ladder.

For buildings four stories or higher, but within reach, aerial ladders or elevating platforms are generally used for rescue. The operator should bring the ladder or platform in above the window, then lower it to the correct height. Otherwise, a panicky victim may try to jump onto the ladder or the platform.

THE HOTEL OR FACTORY RAISE

When victims are trapped on more than one floor, the hotel or factory raise is a way to rescue them with one ladder. Raise an extension ladder close to the building in line with a vertical row of windows. If the ladder has poles, two firefighters must stay in front and use the poles to stabilize the ladder and hold the tip against the building. Guy ropes can be tied to the tip and stretched parallel to the building to prevent lateral movement and add stability.

The heel and beam firefighters anchor the base of the ladder. Then victims can step from the lower windows and climb to the ground on the inside of the ladder. The victims from the upper floor windows can climb down the outside of the ladder.

Firefighters can expect terrified victims to have trouble climbing down the inside of a ladder leaning against a building.

This particular raise is used in critical rescue situations. Firefighters will be busy doing many things and may not be able to individually assist escaping victims who are climbing down the ladder. Therefore, the victims will be greatly assisted if the hotel or factory raise is positioned properly. There are three considerations:

Factory raise considerations

1. The tip of the ladder must be on a windowsill so victims trapped there can escape down the outside of the ladder.
2. The heel of the ladder must be far enough from the building to give victims enough room to climb down the inside of the ladder.
3. The heel of the ladder must not be too far from the building or those trying to use the inside of the ladder will be at a precarious angle and may fall or be unable to climb down.

Table 6.1 shows how the distance from the building to the ladder heel will alter the distance from the upper windows to the ladder. A ladder six feet (2 m) from the building at the heel, with the tip on the fifth-floor windowsill, can be grasped easily from the fourth floor, but will be four feet (1.3 m) from a victim on the second floor. If the heel of a ladder to the fifth floor is three feet (1 m) from the building, the fourth-floor victim would have only seven inches (18 cm) of space in which to climb. That victim could swing around and climb down the ladder from the outside, but if everyone did that with the ladder at such a steep angle, it could be pulled away from the building unless the tip is tied or firefighters hold the poles securely.

TABLE 6.1
Hotel or Factory Raise
Distances of Ladder from Windows at Different Levels
(ladder tip at fifth-floor windowsill)

Distance of Heel from Buildings		Distance of Ladder from Window Second Floor		Third Floor		Fourth Floor	
ft	cm	in	cm	in	cm	in	cm
3	1	25	63	15	38	7	18
4	1.3	33	85	20	51	10	25
5	1.6	40	102	25	63	12	30
6	2	48	122	30	76	15	38

RAISING OR LOWERING VICTIMS ON A STRETCHER

Removing injured victims from elevations above or below ground level requires special consideration. Imprudent techniques have often made the victim's injury worse, so injured persons should be properly secured to some type of stretcher before moving. The stretcher used depends entirely on the equipment available. The possibilities range from an improvised stretcher (Figure 6.7) to stretchers such as a Stokes basket designed for rescue work.

> Firefighters should not be used as "victims" when practicing hoisting and lowering stretchers. Instead, use a mannequin or an empty stretcher.

Figure 6.7 Examples of items that can be used as stretchers if the need arises. *Courtesy of Lisle-Woodridge, Ill., Fire Dist.*

Using a stretcher to move victims reduces additional injury, but increases the rescuer's handling problems. When carrying a stretcher by hand is not practical because of elevation differences, firefighters usually lower the stretcher with a rope and a suspension sling. Typically, sling systems are referred to by the number of points of suspension, such as a "one-point" or a "four-point" suspension system. The number denotes the number of attachment points that determine the stretcher's center of gravity. The most common are the "one-point" and "four-point" suspension systems. The firefighter should not confuse the number assigned to the system with the number of lines attached to the stretcher. Using a one-point system, the line or lines are attached so the center of gravity remains in one position throughout the movement (Figure 6.8). Using a four-point system, the attachment of lines results in four separate effects on the center of gravity (Figure 6.9). The position of the stretcher will depend upon the tension applied to each of the four points. Therefore, the one-point suspension system can be moved with less rocking than can a multiple-point suspension system. Because the stability of the one-point suspension is inherently better, it will receive the emphasis in this section.

Suspension systems or harnesses for a stretcher are generally available from the manufacturer. Typically, these are one-point harnesses with snap-lock connectors for quick and safe attachment and detachment at an emergency scene (shown in Figure 6.8).

A commercially built harness is best for rescue work because it can be easily attached and removed and requires less skill in rope work. But if a commercial harness is not available, a harness can be made with rope and standard fire department knots. Techniques depend on the stretcher used.

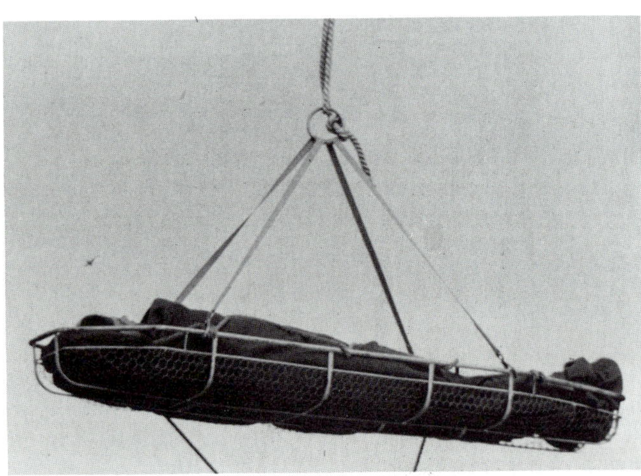

Figure 6.8 With one-point suspension the center of gravity remains in one place throughout the movement of the stretcher. *Courtesy of Montgomery County, Md., Fire/Rescue Services.*

Figure 6.9 Four-point suspension can be useful in situations where the center of gravity must be shifted at some point during the stretcher's travel. *Courtesy of Montgomery County, Md., Fire/Rescue Services.*

ARMY STRETCHER

The army-style stretcher is a cloth stretched tightly between two side beams forced apart by a center brace near each end. The crossbars are hinged so the stretcher can be closed for storage. This stretcher has two handles protruding from each end.

One-point suspension system

To use a one-point suspension harness with an army stretcher to move a victim in a horizontal position takes the following steps:

Step 1: Measure from the end of the rope to 3½ times the length of the stretcher.

Step 2: Fold the running end of the rope back to the standing part to double the measurement.

Step 3: Tie a bowline on a bight and adjust the two separate loops to equal lengths. Place one loop over each set of stretcher handles. Pass the rope between the handles and back over the top of the handles to form small loops around the handles. These loops keep the rope from slipping and the stretcher from tilting inside the suspension harness. Extra loops may be made for extra protection.

To use a one-point suspension harness with an army stretcher to move a victim feet first vertically, pass the working end of the rope through the D-rings at the head of the stretcher and attach it to the handles with a bowline knot and clove hitches. This method will allow victims to be lowered feet first when the only available opening is narrow.

Note

Army or canvas stretchers are notorious for failing under stress. They should not be used for hoisting or lowering a victim unless nothing better is available.

STOKES BASKET

The Stokes basket is a basket stretcher that conforms to the shape of the body. This stretcher provides excellent support for the injured victim and secures the victim in the basket with pre-attached straps. The Stokes is traditionally a metal rod frame with a wire mesh covering, but some of the newer models are molded plastic. Use a Stokes basket by following the steps below:

Step 1: A rope in a one-point suspension system can be used with a Stokes basket to lower the victim in a horizontal position. Place the running part of the rope down through the handle near the victim's feet and push it up through the handle near the victim's left knee. Take the running part over the victim's legs, down through the handle near the victim's right knee, and back toward the head. Push the running part through the handle near the right shoulder and pull it even with the rescuer's shoulder.

Step 2: Pull the rope over the victim's legs to a length equal to the slack brought from the standing part of the rope. Take a bight in the center of this loop, and hold it over the center of the basket.

Step 3: At the head of the stretcher form a loop in the standing part of the rope slightly below the level of the running end. Push the running part through the loop and start a bowline knot. When the running part has passed the loop in the standing part, take it through the loop of the bight from the foot of the stretcher. Pass the running part around the standing part, back through the loop of the bight, and back through the loop of the standing part. Pull the knot tight, and tie one or two half-hitches around the running part for safety. Adjust the suspension harness to distribute the weight evenly.

A one-point suspension for vertical raising or lowering is easily made:

Step 1: Pass the running part down through the right or left handle, under the stretcher, and up through the opposite handle. Attach it to the standing part with a bowline knot. Another method is shown in Figure 6.10.

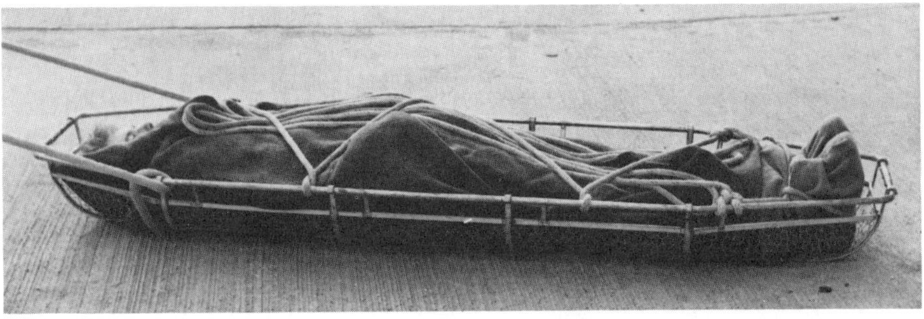

Figure 6.10 Method of attaching hoisting or lowering rope to Stokes basket.

Step 2: Secure the victim in the basket before moving to immobilize injuries and prevent slippage. Use the strapping provided with the stretcher, if available, or rope. Figures 6.11-6.14 illustrate using a rope to secure the victim. If firefighters have to use rope, wrap the victim in blankets, for padding, with arms to the sides.

Step 3: Secure the victim by looping a length of rope equal to the length of stretcher at the end of the running part. (*Note:* If the tie rings are too small, this procedure cannot be done.) Push the running part of the rope through the tie ring near the left or right shoulder. Pull the running part across the victim's shoulders to the tie ring near the opposite shoulder. Then thread the running part shoestring style across the victim to the tie ring near the waist, across the victim to the tie ring near the ankles, across the victim's ankles to the opposite tie ring, back up across the victim to the waist, and back up across the victim to the shoulder tie ring.

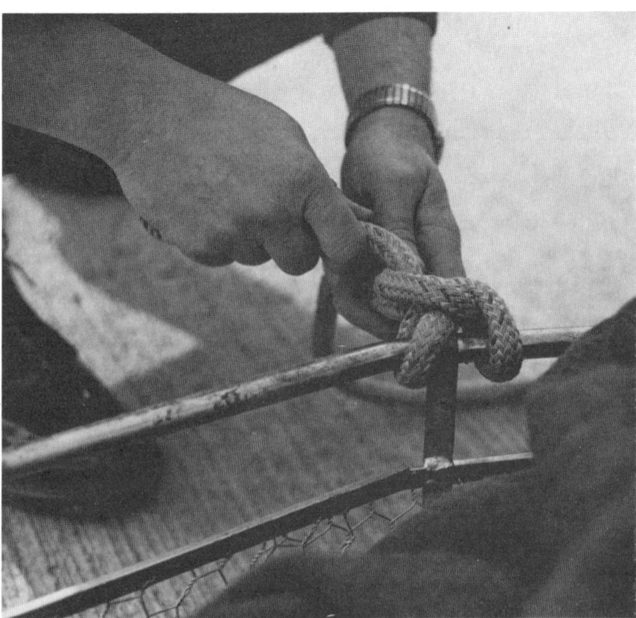

Figures 6.11 and 6.12 Straddle the upright near the victim's shoulder with the first knot.

Figure 6.13 Detail showing securing rope and feet.

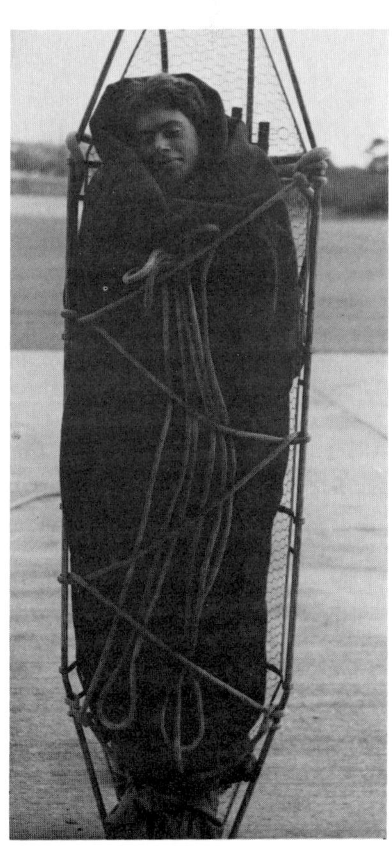

Figure 6.14 The victim secured for hoisting or lowering.

Step 4: Take a bight in the section crossing the victim's chest and ankles. Pull excess slack from the rope. The standing part at the beginning shoulder should be over the top of the tie-ring and secured to the running part. Ideally, this knot will be between the victim's shoulder and the side of the basket.

Step 5: The slack taken at the beginning is now passed through the bight in the section, across the shoulders, and attached with a clove hitch. Maintain a foot of slack between the knot and the bowline at the shoulder. A half hitch may be placed around the two sections of the running part to keep the slack between the shoulder and the bight.

Step 6: Take the running part down to the bight at the ankles and run it through. Pull this line taut to take out the remaining slack and secure the rope with a sheet bend. This will secure the victim and prevent the rope across the shoulder from slipping around the victim's neck. (If the initial slack in the running end cannot be taken, the opposite end of the rope should be used for the last adjustment.)

The head may need to be stabilized further with a cervical collar and head strapping.

STRENGTHENING A STOKES BASKET

Stokes basket may need reinforcement

The butt weld holding together the tubular metal frame of Stokes baskets has been known to fail under stress. To prevent this is a simple task (Figures 6.15 and 6.16).

1. Dress the factory weld by grinding it flush with the tubing surface.
2. Clean and sand the tubular frame between its supports to remove paint, rust, and other foreign matter.
3. Cut a length of 16-gauge steel tubing 3/4-inch (20 mm) in diameter to fit between the supports, then cut this tubing lengthwise.
4. Drill a small hole (about 1/16-inch (1.5 mm) diameter) in the tubular frame as a vent for built-up air pressure caused by the heat of welding.
5. Weld the two halves of the tubing over the tubular frame as shown in Figure 6.16. Be sure that the longitudinal joints are perpendicular to the plane of the stretcher bottom.
6. Clean and paint the area to prevent corrosion.

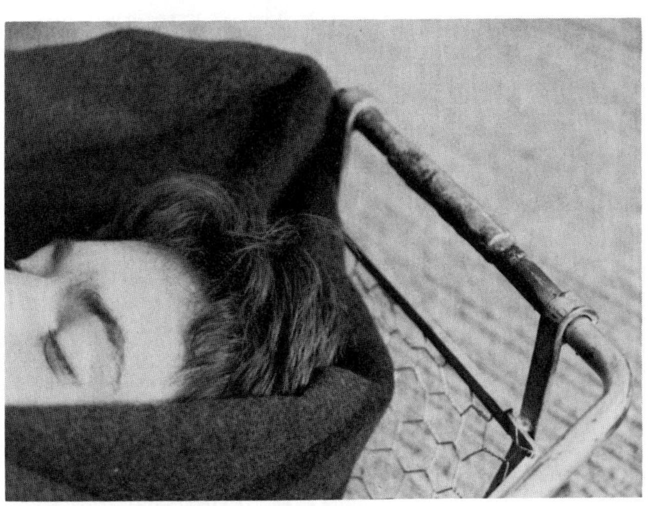

Figure 6.15 The factory butt weld of a Stokes basket should be reinforced to prevent failure under stress.

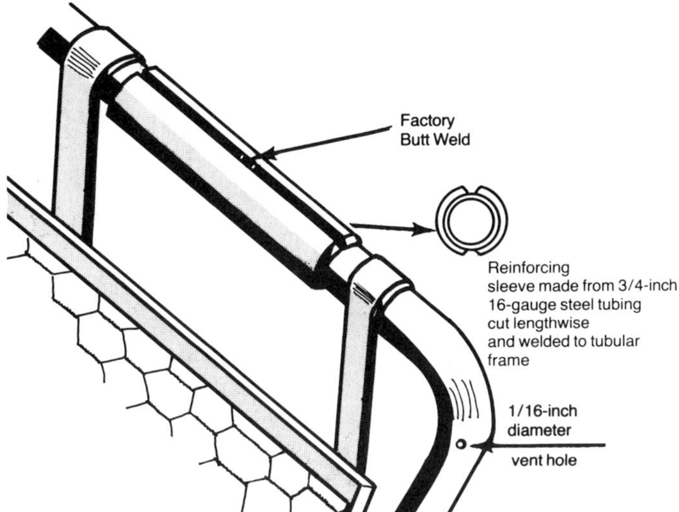

Figure 6.16 Detail drawing of how to reinforce the Stokes basket weld.

FLAT BACKBOARDS

Flat backboards are stretcher-like devices that provide a way to immobilize a victim for transport. They are typically built of rigid material 16 to 24 inches (40 to 60 cm) wide, and 72 to 81 inches (182 to 205 cm) long and have the same basic strap configuration as Stokes baskets. The same method of securing and suspending the victim may be used. There is a modified flat backboard called a half-back, which is 32 to 40 inches long (81 to 101 cm). A half-back is only acceptable for raising and lowering small children.

WRAPPING A VICTIM BEFORE SECURING TO A STRETCHER

A victim may have to be wrapped inside blankets before transporting on a stretcher to combat shock by keeping the victim warm and to provide better padding and support for lashing the victim to the stretcher. Follow the steps shown in Figures 6.17-6.20.

Step 1: Use two blankets. Place the first on the head of the stretcher with the length perpendicular to the stretcher. About 12 inches (30 cm) of blanket should hang over the stretcher head. Place the second blanket on the foot of the stretcher with its length in line with the stretcher. About two feet (60 cm) of blanket should hang over the foot of the stretcher (Figure 6.17).

Step 2: Fold or roll the excess blanket at the head of the stretcher into a small pillow for the victim's head. (*Note:* If the stretcher has head harness, fold the blanket down to the point just below the harness attachments.) Gently lay the victim on the stretcher with arms to the sides and feet together (Figure 6.18). Fold the corners of the blanket, longer side first, over the victim, and strap the victim in (Figures 6.19 and 6.20).

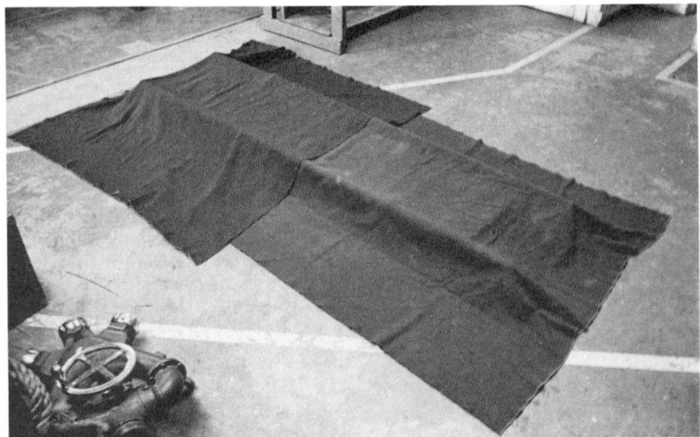

Figure 6.17 Lay two blankets over the stretcher as shown here.

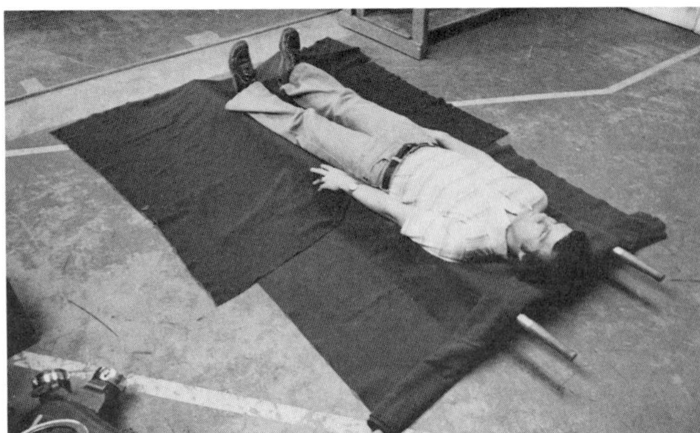

Figure 6.18 Fold the head blanket overhang into a pillow and place the victim on the stretcher.

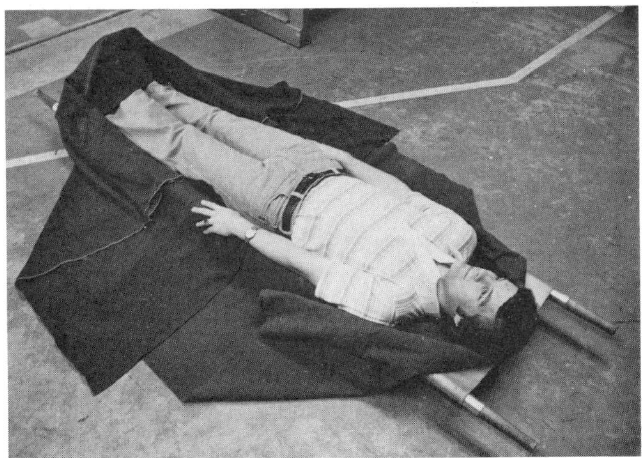

Figure 6.19 Fold the foot blanket over the victim's feet and fold the corner in to the victim's sides as shown.

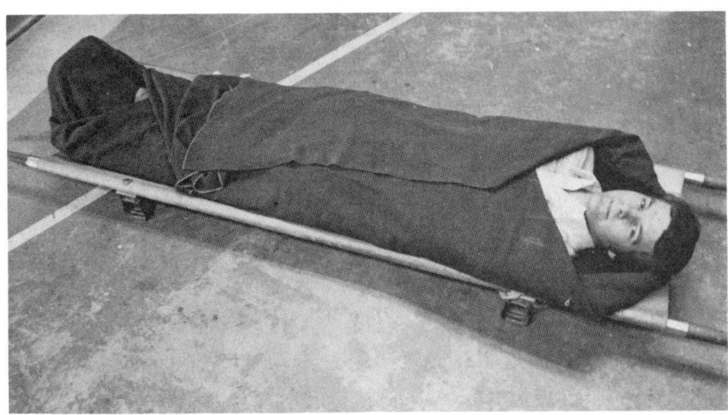

Figure 6.20 Fold the sides of the blankets over the victim to complete the wrap.

Step 3: The victim's head should be immobilized as explained in IFSTA's **Fire Service First Aid Practices.** If using a backboard, put a suitable object underneath at each end before lashing the victim in so firefighters can reach under the stretcher. If nothing else is handy, a firefighter's boot laid on its side will work.

Step 4: At least two people are needed to secure the victim. Pull 35 feet (10.5 m) off the running end of a rope and attach the standing part to the stretcher handle by the right shoulder for an army stretcher, or the hand hole of a backboard, with a clove hitch and a safety (Figure 6.21). Pass the running part over the victim's chest, back under the stretcher, and through the standing part to form a hitch (Figure 6.22). Adjust the slack in the rope until the line crosses the victim just above the elbow and the hitch lies just below the right breast. Lash a second hitch around the victim the same way. It should cross the victim just below the groin and rest over the right thigh (Figure 6.23). Lash a third hitch around the victim, crossing at mid-shin (Figure 6.24).

Figure 6.21 With one end of the rope put a clove hitch and safety over the handle near the victim's right shoulder.

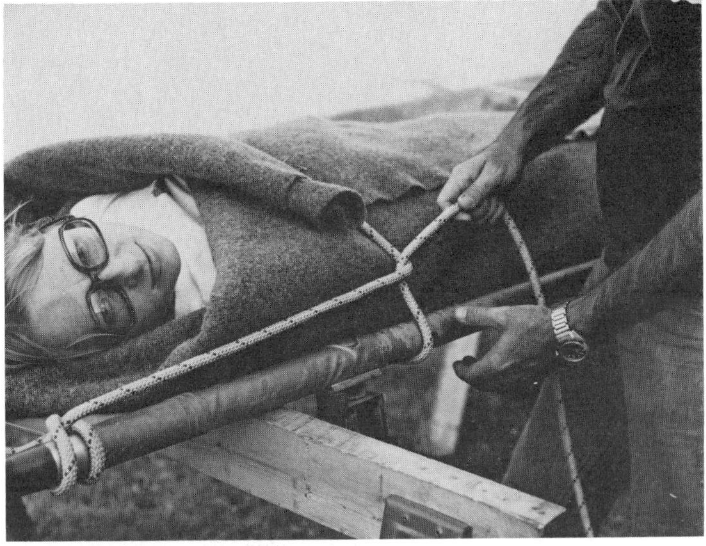

Figure 6.22 Pass the running part of the rope over the victim's chest, under the stretcher, and around the standing part to make a hitch.

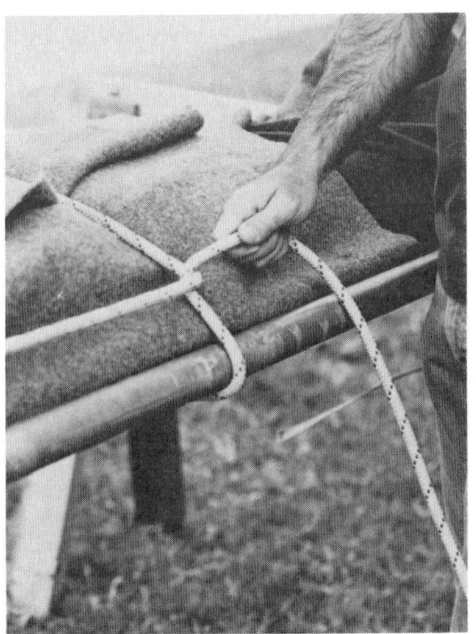

Figure 6.23 Do the same at the thighs . . .

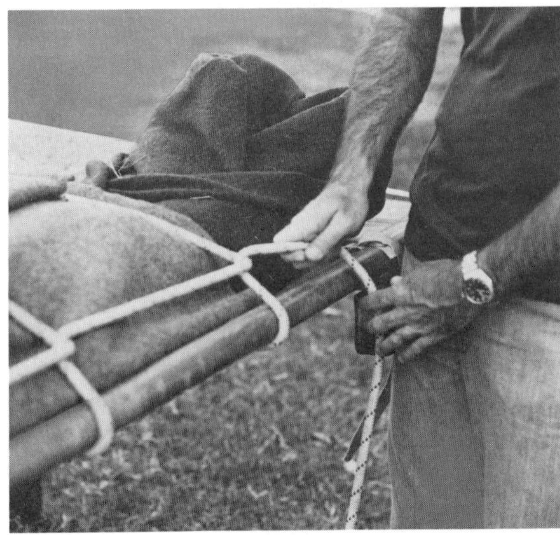

Figure 6.24 . . . and at the shins.

Step 5: Tighten the victim into the lashing by grasping the rope below the first hitch and easing it toward the right edge of the stretcher. With the victim snugly held, the second firefighter can secure the hitch by pressing on the bight with thumb and forefinger. Adjust the slack through the second hitch, and repeat the procedure. Tighten the final hitch using the thumb-forefinger technique.

Step 6: When all the hitches have been tightened, pull the running part of the rope past the foot of the stretcher while someone holds the last hitch.

Step 7: The firefighter tying faces the stretcher and forms a loop as shown in Figure 6.25.

Figure 6.25 Form a loop with the running part . . .

Step 8: Place the loop gently over the feet and gently pull the running part of the rope snug (Figure 6.26). Tighten the rope for proper tension, bring the rope around the feet (Figure 6.27), and form a hitch as shown in Figure 6.28.

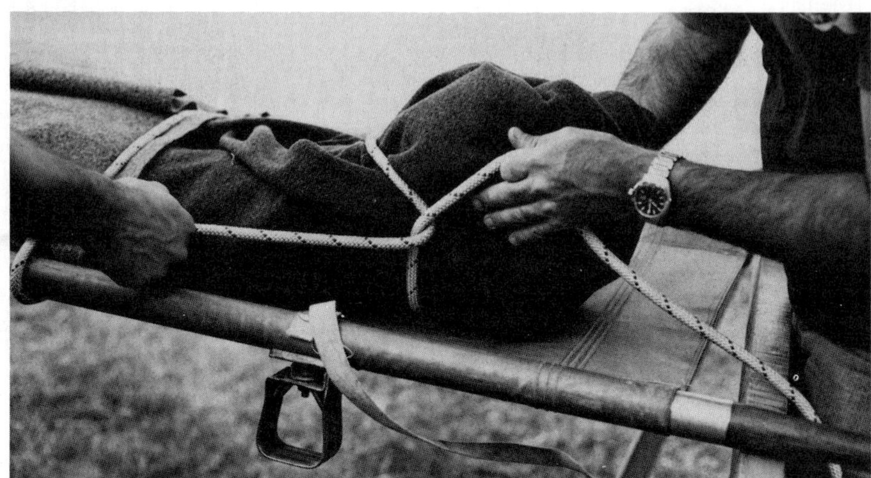

Figure 6.26 . . . and tighten it around the victim's ankles.

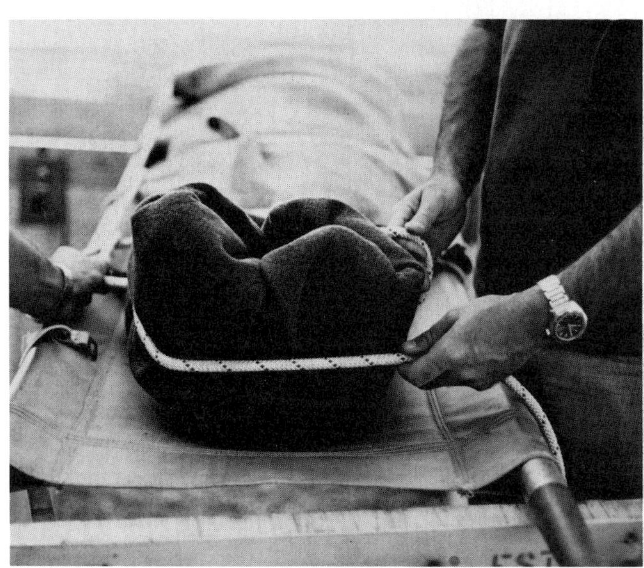

Figure 6.27 Bring the running part across the victim's soles . . .

Figure 6.28 . . . and run it through the ankle loop to form a hitch.

Step 9: Take the running part of the rope toward the head, along the left side of the stretcher. At the loop over the shins, turn the running part past the loop and back to the foot of the stretcher. Run the rope between the loop and the leg and back toward the head to form a bight around the loop (Figures 6.29 and 6.30). Pull the slack out of the rope and apply the proper tension. Move the thumb-forefinger tension hold from the feet to the newly formed hitch. Repeat this procedure at each of the remaining loops along the left side of the stretcher (Figures 6.31 and 6.32). When the final hitch has been pulled snug and the hitch is being held securely, attach the running part of the rope to the left stretcher handle or the backboard hole with a clove hitch and a safety (Figures 6.33-6.34).

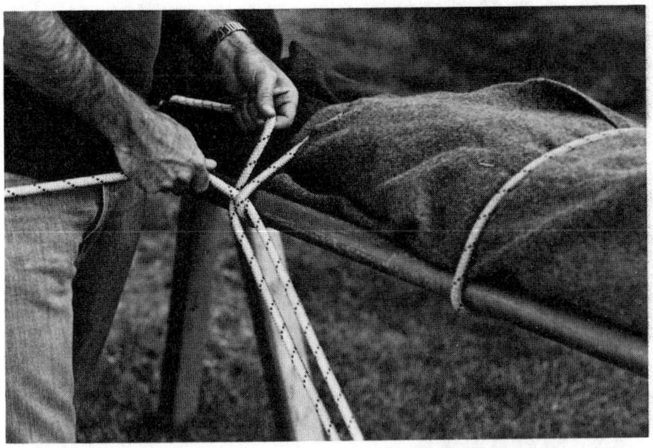

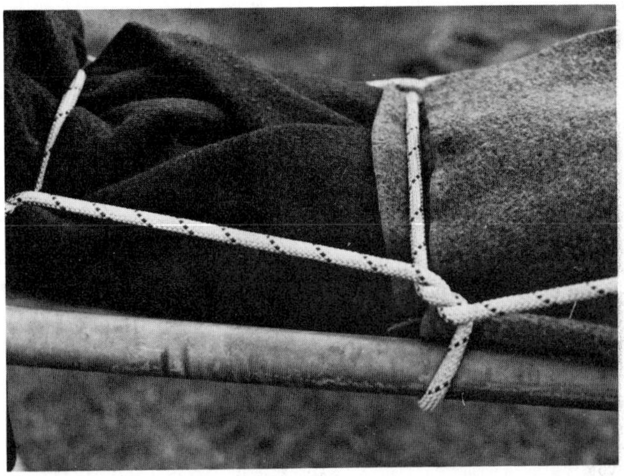

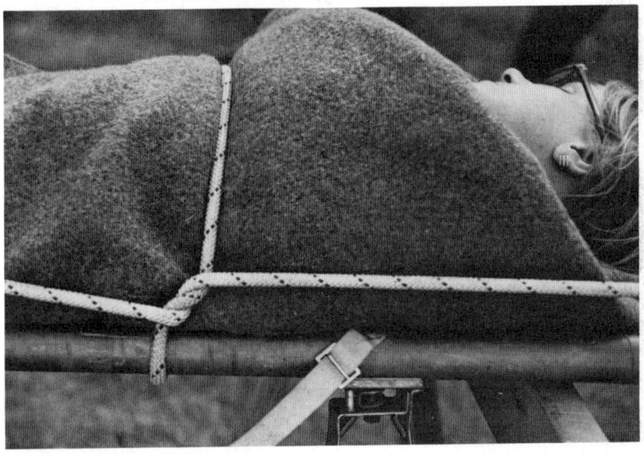

Figures 6.29-6.32 Form hitches with the running part through the loops over the victim's shins, thighs, and chest.

Figures 6.33 and 6.34 Secure the running part over the left handle with a clove hitch and safety.

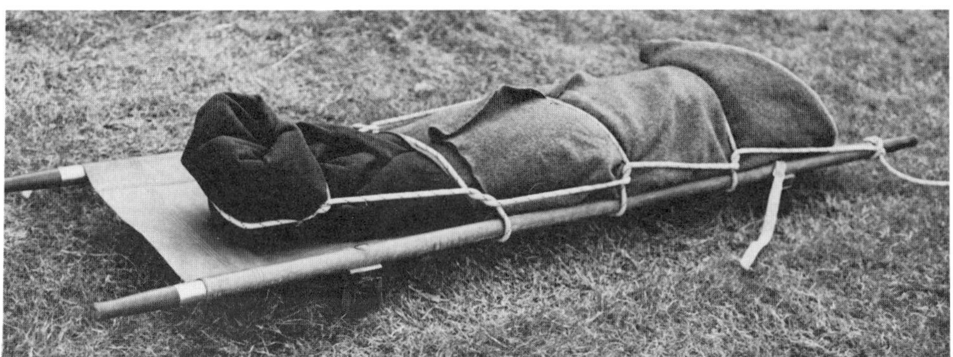

Figure 6.35 The victim as wrapped and secured for transport.

Step 10: Use this lashing technique to move the victim when the feet are below the head. It works because the weight of the victim is transmitted to the hitch around the feet and keeps the lash snug. This makes an excellent lashing technique for lowering the victim in a vertical position. However, when the victim is in a horizontal or shock position, the lashing may loosen if the victim shifts to the head of the stretcher. When the victim is transported in a position other than vertical, apply a tension strap to the loop around the feet. A tension strap is tied as follows: Use a short section of rope attached through the loop around the feet by forming a bight in the rope and passing it between the soles of the feet and the lashing around the feet. Pull the two ends of the short rope through the bight, pull them tight, and attach them to the handles at the foot of the stretcher or the hand holes at the foot of the backboard.

IMPROVISED STRETCHERS

Turnout Coats and Pike Poles

Satisfactory stretchers or litters can be improvised with two turnout coats and two pike poles. The coats can be slipped onto the poles easily by turning them inside out over the head of a person who is holding the two poles. Turn the flaps around the poles and button or snap underneath.

Salvage Covers or Blanket and Pike Poles

A stretcher can be improvised with two pike poles or similar equipment and a blanket or salvage cover. When using a salvage cover follow the procedures below.

Step 1: Fold the salvage cover in half lengthwise (Figure 6.36).

Step 2: Fold under about 18 inches (45 cm) at one end (Figure 6.37).

Figure 6.36 To make a stretcher from a salvage cover and two pike poles, fold the salvage cover in half lengthwise. *Courtesy of Harold Prevost.*

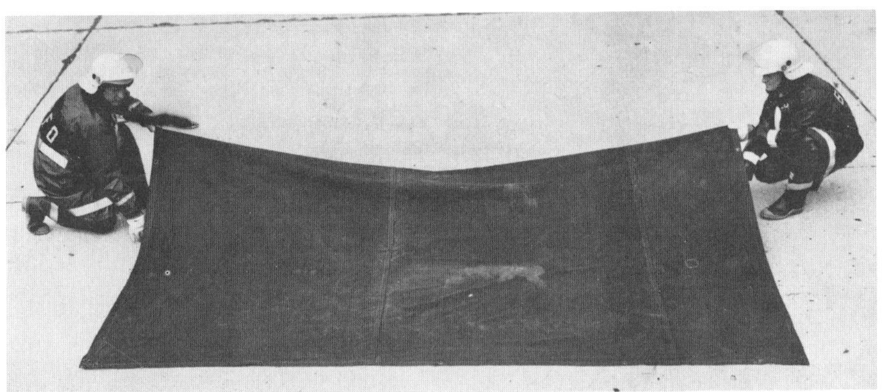

Figure 6.37 Fold under about 18 inches at one end. *Courtesy of Harold Prevost.*

Step 3: Place the first pike pole across the salvage cover in the same direction as the original fold, about one foot (30 m) from the center (Figure 6.38).

Figure 6.38 Place first pike pole about one foot from center on top of salvage cover. *Courtesy of Harold Prevost.*

Step 4: Fold the shorter side of the salvage cover over the first pike pole (Figure 6.39). *Note:* Placing the inside foot on the pike pole prevents the pole from slipping out of position while the fold is being made.

Step 5: Place the second pike pole on top of the salvage cover, parallel to and about two feet (60 cm) away from the first pike pole (Figure 6.40).

Step 6: Fold the top layer of the salvage cover over the poles (Figure 6.41). *Note:* Failure to do this step before Step 7 could allow the cover to slip when victim is placed on the stretcher.

Figure 6.39 Fold shorter side of salvage cover over the pole. *Courtesy of Harold Prevost.*

Figure 6.40 Place second pike pole about two feet from the first pike pole. *Courtesy of Harold Prevost.*

Figure 6.41 Fold shorter (top) side of salvage cover over the poles. *Courtesy of Harold Prevost.*

Step 7: Fold the long side of the salvage cover over the poles (Figures 6.42 and 6.43).

Step 8: Place the victim on the stretcher with feet under the fold formed in Step 2 and fold salvage cover over the victim (Figure 6.44).

Step 9: Carry the stretcher by lifting each end of the pike poles (Figure 6.45).

Figures 6.42 and 6.43 Fold the rest of the salvage cover over the poles. *Courtesy of Harold Prevost.*

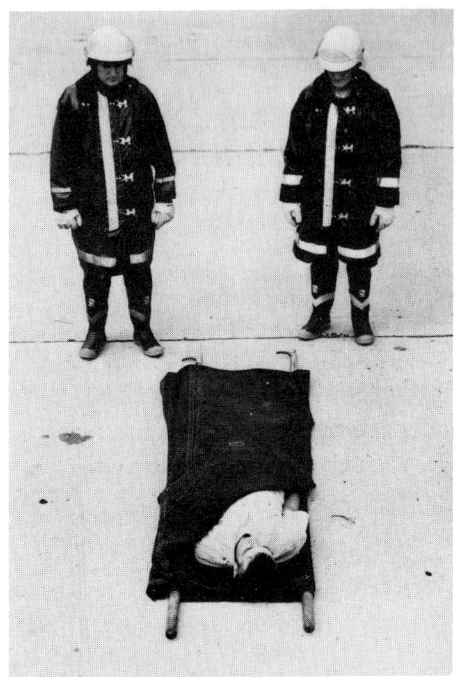

Figure 6.44 Place victim on stretcher, feet under the fold formed in Step 2 (Figure 6.37), and fold the salvage cover over the victim. *Courtesy of Harold Prevost.*

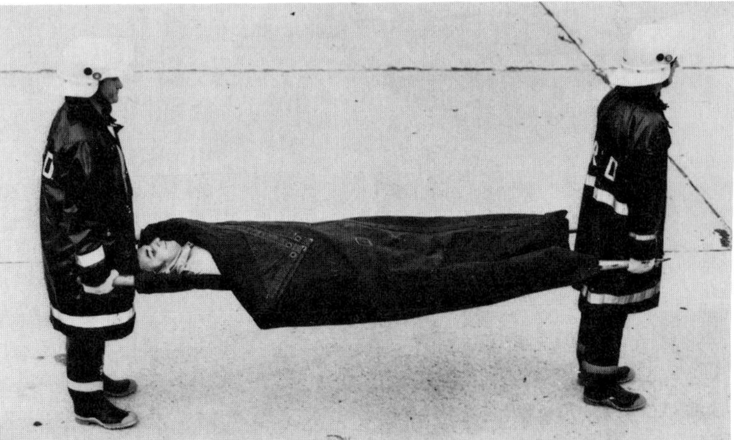

Figure 6.45 The victim is now ready for transport. *Courtesy of Harold Prevost.*

When using a blanket, place one pole on the unfolded material about one foot (30 cm) from its center, and fold the short side of the material over the pole toward the other side. Place the second pole on the two thicknesses, about two feet (60 cm) from the other pole (Figure 6.46), and fold the other side of the material over the second pole toward the first. When an injured person is placed on the blanket, the folds are held together by the weight of the body.

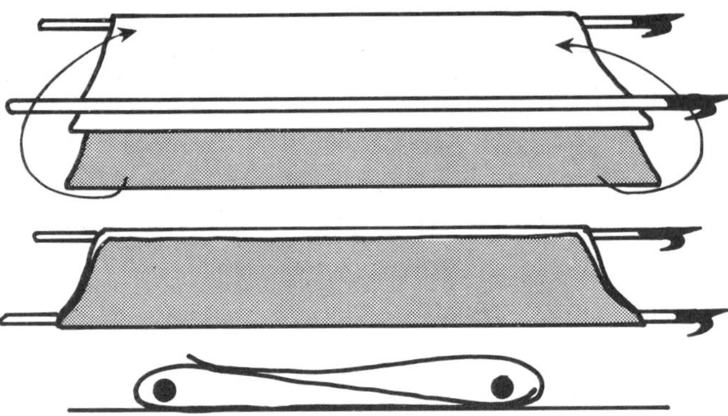

Figure 6.46 Making an improvised stretcher from a blanket and two pike poles.

LADDER FULCRUM

When a victim on a stretcher has to be lowered from a place that can be reached with a single wall ladder, the ladder itself is often used as a fulcrum and lowering device. At least four firefighters are needed, three at the top of the ladder and one on the ground. One rope is needed.

Step 1: Position the ladder at a correct climbing angle at the victim's location. Two firefighters balance the stretcher on the building edge while the third lashes the head of the stretcher securely to the top rung of the ladder. The lashing procedure will depend on the type of stretcher.

Step 2: Secure the head of the stretcher so the ladder supports most of the victim's weight (Figure 6.47). After lashing the head of the stretcher, attach the rope to the foot of the stretcher. If possible, the firefighter tying the ropes takes the standing part of the rope about 20 feet (6 m) away from the edge. This firefighter is the anchor for lowering (Figure 6.48).

Step 3: The firefighter on the ground pushes the ladder toward the building, raising the head of the stretcher. When the ladder is adjacent to the building, or at least within one stretcher length, the firefighter on the ground heels the ladder from the side away from the building. The two firefighters holding the stretcher push it out the window as the anchor firefighter slowly feeds out rope. When the stretcher is out the window, the firefighter on the ground gently guides the ladder away from the building as the anchor and the other two firefighters keep enough tension on the rope to keep the stretcher level (Figures 6.49 and 6.50).

Step 4: When the stretcher is grounded, remove it and raise the ladder so the firefighters can descend.

Figure 6.47 Using a ladder as a fulcrum: The stretcher is securely attached to top rung and a guide rope attached to the stretcher's foot. *Courtesy of Lisle-Woodridge, Ill., Fire Dist.*

Figure 6.48 The two ends of the guide rope are held by firefighters about 20 feet from the edge. *Courtesy of Lisle-Woodridge, Ill., Fire Dist.*

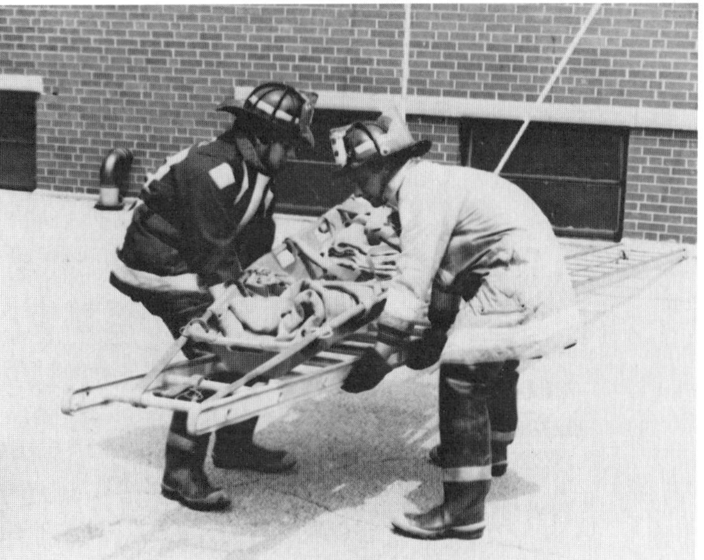

Figures 6.49 and 6.50 The ladder and stretcher are lowered gently, keeping the stretcher as horizontal as possible. *Courtesy of Lisle-Woodridge, Ill., Fire Dist.*

GUIDELINES

After the victim has been securely lashed to the stretcher and the actual movement begins, a guideline is necessary for rescuers to control the swing of the stretcher or to keep it away from walls or other obstacles.

Stretcher position will determine where the guidelines should be attached. If the stretcher is being lowered in a horizontal position, attach the guidelines at the head and foot of the stretcher. If the stretcher is being lowered in a vertical position, attach the guidelines to each side.

Firefighter positioning important

The position of the firefighters handling the guidelines from below is important. The closer they stand to the stretcher's destination, the more their pull will be transferred to the persons holding the weight. The firefighters controlling the guidelines should be to the side of and out from the stretcher destination. This gives them additional control without adding additional weight to the load.

SLIDING A STRETCHER DOWN A LADDER

A victim on a stretcher can be slid down a ladder without much trouble. The technique involves running the handles of common tools, such as pike poles and shovels, through the bottom "D" rings of army-style stretchers or lashing these tools across the bottom of a Stokes basket or backboard from hand hole to hand hole. Secure a rope to the head of the stretcher with a bowline.

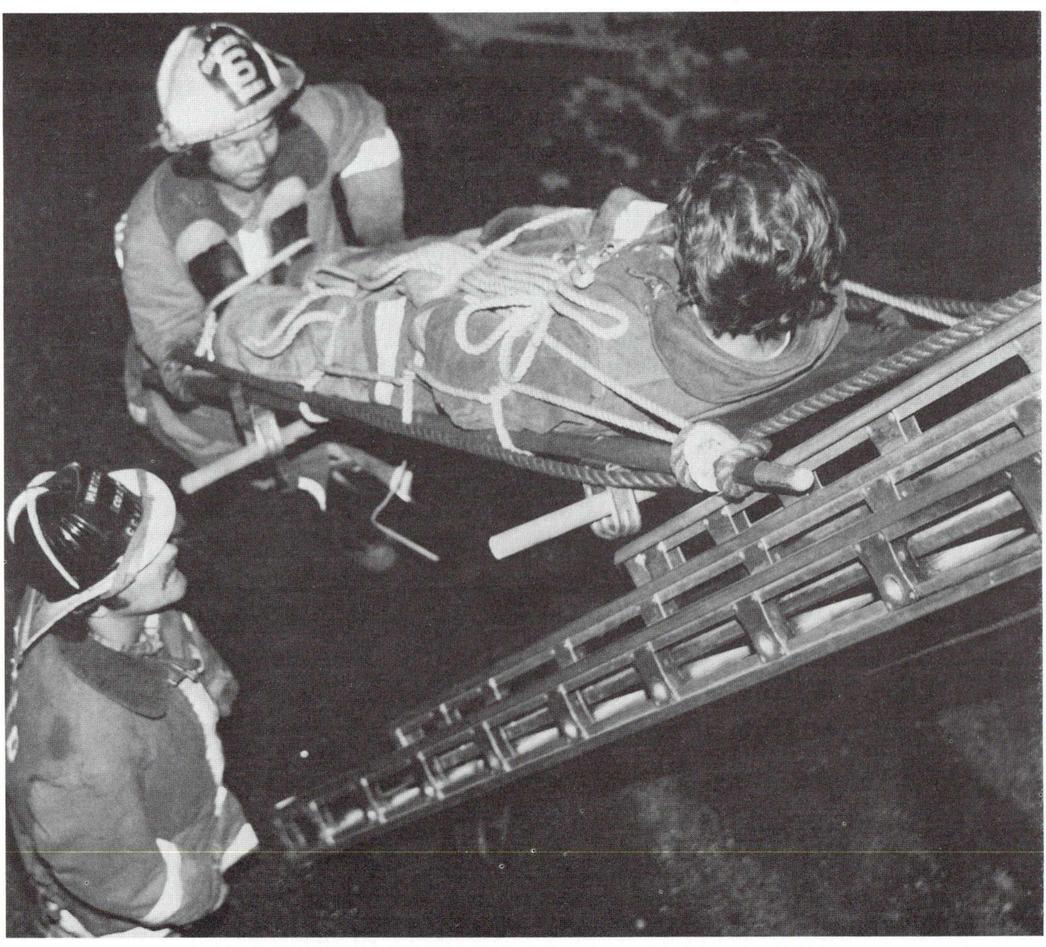

Figure 6.51 Sliding a stretcher down a ladder. Note safety line attached to stretcher head. *Courtesy of Bill Maggi.*

Adjust the ladder so the tip is level with the window ledge. One firefighter climbs onto the ladder to guide the stretcher when it is placed on the ladder. Another firefighter anchors the rope about 20 feet (6 m) from the window by taking the rope around the waist, slightly below the buttocks, and backs over the standing part to form a bight. Two firefighters then gently slide the stretcher onto the ladder as the anchor keeps tension. When the stretcher is on the ladder, the two firefighters help the anchor lower the stretcher as it is guided by the firefighter on the ladder (Figure 6.51).

JUMPERS

Firefighters and rescue personnel are sometimes faced with the difficult situation of assisting individuals who have accidentally placed their lives in jeopardy or have been forced into such a situation by fire. These situations are often in areas high above the ground or water. Office workers and apartment dwellers have been forced to windows, balconies, or ledges by heat, smoke, and fire in their offices and living areas.

Some distraught individuals seek a location from which to jump, such as building windows, outside ledges, roofs, and rotundas inside buildings. In addition, bridges over water, roadways, or structures within industrial complexes have also been used as jumping sites.

Instructions from rescuers, especially in situations where the victim is still in danger, should be given over public address systems or building systems, if available. Rescuers should repeat and update instructions to those waiting for help because any movement could further jeopardize a victim's position. Continued contact with rescuers assures victims that someone knows of their situation and that help is on the way.

Assure victim often

Suicide Attempts

Special precautions must be taken with the individual who is threatening to jump as a means of suicide. *No person threatening to jump should ever be taken lightly.* Usually, the overall situation will be handled by police, but the firefighter may be involved. Often fire and police department chaplains have played an important role in getting people to change their minds or at least in comforting them until additional aid can be obtained.

Often a firefighter or rescue squad member can more easily gain the confidence of a jumper easier than others can because firefighters want to help those in trouble and they do it daily.

If other ways to rescue the victim have failed, the final chances of saving the victim may be the life net or air bag.

LIFE NET PRACTICES

Many things can happen when jumping into a life net and its use is extremely hazardous, but they have been used for some successful rescues that could not have been made otherwise.

Open net away from use location

A life net should be opened and firefighters positioned around it at a safe distance from where it will be used. If it is carried to the point of use before it is completely ready, an excited victim may jump before the firefighters are ready. The most practical way to carry a one-quarter-fold net is vertically with the frame side down. Four firefighters should carry the net and open it. Carry the net by the frame, not the springs (Figure 6.52).

The numbers of the positions are 1, 2, 3, and 4, designating the part that each firefighter in that particular position must execute. Firefighters should be able to handle any of the positions and do the part required. Many departments find it practical to paint the top frame of the open end fluorescent or white, which avoids confusion and facilitates uniform operation. At a safe distance from the place of service, place the one-quarter-fold life net on the ground horizontally with the folded fabric side up. The steps for opening are described below.

Step 1: No. 2 steps back to the middle of the net on the fabric side. No. 1 stands at the open end and No. 3 at the hinged end. No. 4 reaches across the net and grasps the two top rails of the frame near the open end. Each firefighter can open the net to one-half fold size (Figure 6.53).

Step 2: No. 4 continues to unfold the frame by pulling the two top rails over. No. 3 checks to see that the hinge locks snap into place (Figure 6.54).

Step 3: No. 2 reaches over and gets the top frame at the hinge from No. 3. No. 2 steps backward to open the net fully (Figure 6.55).

Step 4: The net now opens into a full circle and the hinge locks are snapped into place by Nos. 1 and 4 (Figure 6.56). The four who opened the net are now located at the locks (Figure 6.57). All additional firefighters should take evenly spaced positions between them.

Person may jump anytime — keep watch

The life net is now ready to move into place. This should be done quickly and the handlers should always be ready for the jump in case the victim will not wait for instructions. No fewer than ten firefighters, preferably twelve, should hold a life net when a jump is made.

During life net training, firefighters should learn the proper way to handle and hold a life net. This part of training should be

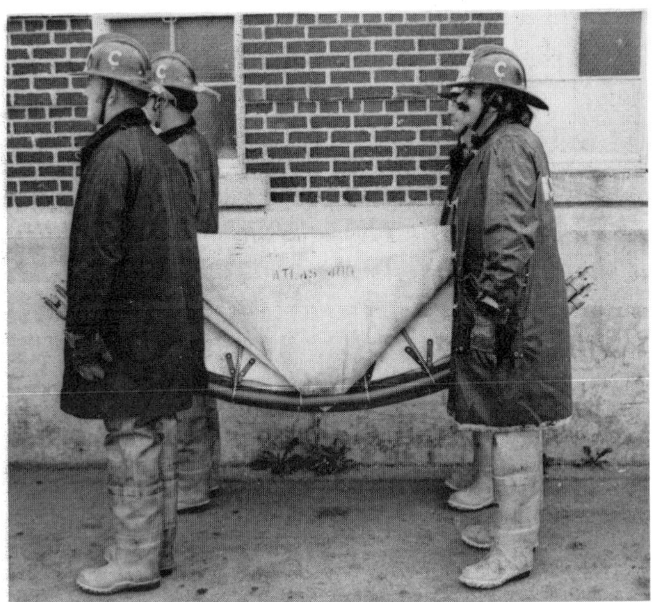

Figure 6.52 Four firefighters carry the life net by the frame. *Courtesy of Springfield, Mo., Fire Dept.*

Figure 6.53 The net is laid on the ground and opened to half-size. *Courtesy of Springfield, Mo., Fire Dept.*

Figure 6.54 No. 3 checks to see that the hinge locks snap into place. *Courtesy of Springfield, Mo., Fire Dept.*

Figure 6.55 The net is opened fully. *Courtesy of Springfield, Mo., Fire Dept.*

Figure 6.56 Nos. 1 and 4 snap the hinge locks into place. *Courtesy of Springfield, Mo., Fire Dept.*

Figure 6.57 The four firefighters who opened the net stand at the locks. *Courtesy of Springfield, Mo., Fire Dept.*

carefully and thoroughly conducted so firefighters will be able to catch a falling body in the center of the net regardless of how a jump or fall is made.

Each firefighter faces the net so that all can move forward, backward, or sideways quickly with absolute freedom. *Grip the frame of the net with the palms up and thumbs underneath the rail. Hold the net about shoulder high with the arms to the side of the body. Otherwise a sudden impact may jam the elbows into the stomach (Figure 6.58).*

It is important for everyone holding the net to watch the jumper. No two people will ever jump alike and the net holders must move quickly to position the net beneath the jumper. Time may not permit verbal direction. The net holders must be able to react as a team to insure proper positioning.

Figure 6.58 Additional firefighters stand evenly between the original four. *Courtesy of Springfield, Mo., Fire Dept.*

Training

When life net training first starts, arrange jumps from a height of about ten feet (3 m) until firefighters get used to the techniques. Jumps can be gradually increased to greater heights as skill is perfected. After a command from the instructor, a jumper should step, not jump, from the platform and raise both legs and arms horizontally while descending to strike the net in a sitting position as shown in Figures 6.59 and 6.60. The net should be cleared for the next jumper. Life nets should be inspected regularly according to manufacturer's recommendations and specifications.

Figures 6.59 and 6.60 Jumper should land in a sitting position on the net. *Courtesy of Springfield, Mo., Fire Dept.*

AIR SAFETY CUSHION

Because of a variety of problems, life nets are rarely used anymore. They are not reliable at the greater heights of many buildings today, and jumpers cannot be instructed how to jump the best way. They may jump too soon or land on their heads, or the rescuers may be injured from improper holding methods. The life net is carried on some ladder apparatus, but the person most likely to successfully jump into it is the trained firefighter.

A new device is available to increase the chances of a successful jump from up to ten stories high, an air-inflated cushion stored in a compact unit (Figure 6.61).

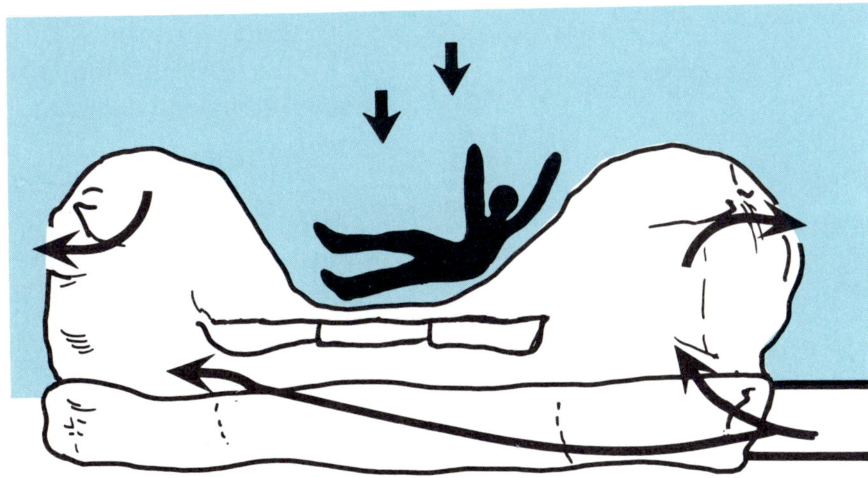

Figure 6.61 Much safer than the life net is the air bag. *Courtesy of Safety Air Cushion, Inc., Metairie, LA.*

The air cushion is designed so that any object landing on the cushion is cradled and held, rather than bounced. This means that two or more falling objects will not collide or be thrown from the bag after a jump. Highly pressurized air is not used, so a small cut is not critical.

The air cushion bag has two air cells. The lower cell is airtight and absorbs twice the kinetic energy absorbed by the upper cell. A special air release system in the upper cell lets some of the compressed air vent upon impact to expel the excess pressure. Fans connected to the apparatus or to a portable generator continuously provide replacement air for the cell. Jumps can be made every five to seven seconds and the falling energy will still be dissipated evenly. The air cushion bag can even be set up on parked cars or other obstructions.

LIFELINE PRACTICES

Sometimes firefighters may be able to reach a floor or roof above a victim, lower themselves to the desired level, and lower a victim with them to the ground. But lifeline rescue is basically used to rescue firefighters. It is not practical for rescuing victims who are not trained in its use, because lifeline rescue is difficult and hazardous. Lifelines must be dependable at all times. Their size and material should be able to stand at least ten times the strain likely to be placed on them during rescue work.

A typical lifebelt (also known as a pompier or safety belt) with provisions for carrying small tools is shown in Figure 6.62. This safety and rescue device should be carefully inspected regularly, including a close examination of the snap and buckles. The belt described here is to be used right-handed, and a left-handed person should adapt to its use with extreme caution. Put the belt on so the back of the hook will be toward the hand controlling the rope. Buckle the belt snugly with equal tension on all straps, and secure all strap ends in the keepers.

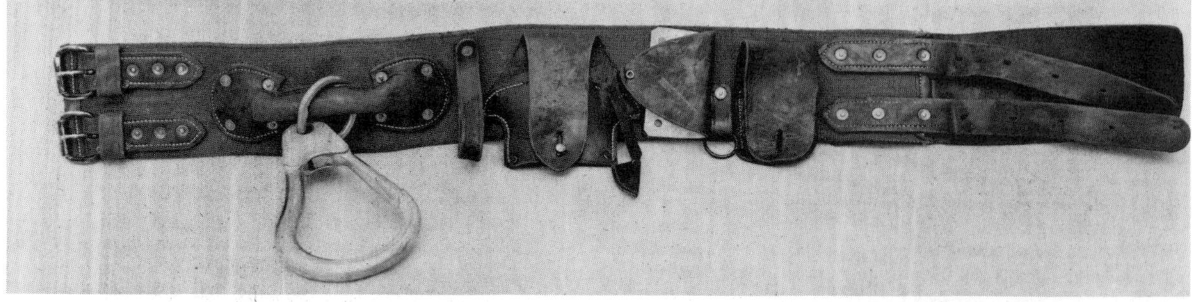

Figure 6.62 A typical lifebelt. Newer models have locks on the snap for greater safety and are highly recommended.

A person who is going to slide a lifeline must always wear gloves. With the rope secured at some point above the rescuer, hold the lifeline along the closed side of the hook with the right hand (Figure 6.63) and throw a loop through the snap with your left hand (Figure 6.64). Continue to hold the hook with your right hand and throw a second loop through the snap (Figure 6.65). Raise the hook on the line as far as possible. Grasp the line high above the hook with the left hand. Pull the rope across the back of the hips with the right hand, ready for the slide. Always apply body weight to the lifeline before attempting to slide the rope.

Figure 6.63 Hold the lifeline along the closed side of the hook with the right hand.

Figure 6.64 Throw a loop through the snap.

Figure 6.65 Throw a second loop through the snap.

Many firefighters have been injured, and some killed, because simple, basic safety measures have not been taken while rappelling. There should be a person stationed on the ground to help stop the rappeller's descent if there is trouble. The person on the ground need only pull down strongly on the rope, tightening the loops around the hook, which will stop the rappeller (Figure 6.66). This safety person must keep watch on the rappeller at *all* times—a second's inattention can be fatal.

A second lifebelt can be used in a rescue. After securing yourself to the line, take an extra turn on the hook to manage the added weight. The victim sits on the windowsill between the rescuer's feet (Figure 6.67). The rescuer drops to the lap of the victim, who snaps the belt into the eye of the snap on the rescuer's belt and puts the arms around the rescuer. The rescuer pushes both clear from the sill with the feet and descends slowly (Figure 6.68).

Figure 6.66 A person must be stationed on the ground to help prevent mishap—especially in other-than-emergency situations. The person on the ground need only pull firmly down on the rope to stop the rappeller's descent.

Figure 6.67 The person to be rescued sits facing the rescuer and between the rescuer's legs. The two belts are snapped together.

Figure 6.68 The rescuer pushes off with both feet and descends slowly.

USE OF THE SWISS SEAT FOR RAPPELLING

Although the lifebelt has been used in the fire service for rappelling for many years, its acceptability for that purpose has recently been questioned because:

- No acceptable means of determining the condition of the lifebelt, to indicate its inservice acceptability for rappelling, has been developed.

- When used for rappelling, the lifebelt tends to shift up to the mid-torso, applying the load to the lower back and rib cage.

Belt Testing

Testing the lifebelt for rappelling is important because of its length and type of service and the stress loads during rappelling. New lifebelts are usually acceptable for stress loads like those encountered in rappelling, but during normal use in fire fighting these lifebelts are exposed to water, dirt, acids, caustics, salts, and adverse environmental conditions. Therefore, it is important that the condition of the lifebelts be determined before rappelling.

Lifebelt testing a "Catch-22"

In addition to the normal wear and tear on the lifebelts, one must consider the stresses that rappelling puts on a lifebelt. A common misconception about this stress is that it is equal to the weight of the person rappelling. The stress is the product of the weight of the person and the gravity-caused acceleration of this weight. This effect of weight and acceleration can create stress loads considerably greater than the weight of the person. For instance, a man weighing 170 pounds (77 kg) can create stress loads on the rappelling equipment near 1200 pounds (544 kg) during a rapid descent. Because of these two factors a method to test a lifebelt to determine its condition is necessary. Unfortunately, the current methods used to test a lifebelt's condition overstress the belt and make it unfit for service. This is a "Catch 22" situation: The belt must be tested to determine if it is in good enough condition to be used; but if it is tested with current procedures it cannot be used in service, because it has been overstressed.

Belt Position

The effect of the belt sliding up the torso causes additional problems. The initial result of this shift is that the weight of the person is concentrated at a point well above the natural center of gravity. When this happens the person must assume a nearly horizontal position to keep proper footing on the rappelling surface.

The shift in position because of the belt slipping up the torso results in the potentially dangerous situation of the stress load being carried by the lower back and rib cage. Physiologically the skeletal structure is intended to carry weight applied down from the upper body and in line with the body plane. In the position that results from the lifebelt slipping up the torso, the load is applied vertically to this plane. Not only is an extreme load applied to one of the less muscular portions of the body but it is applied completely differently from the normal loading on that section of the body (Figure 6.69). Compounding this loading problem is the fact that the two bulkiest portions of the body, the upper torso and legs, act like weighted levers and increase the adverse effect on the lower back (Figure 6.70).

Normal
Loading

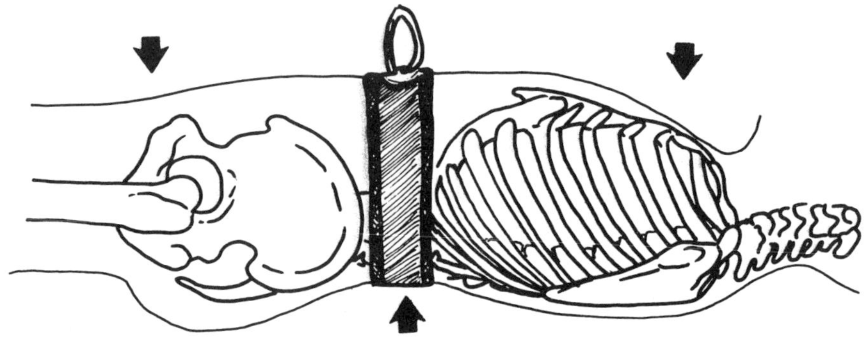

Figure 6.69 The stress load on the skeletal structure when a person rappels with a lifebelt is at a right angle to normal body loading.

Figure 6.70 Compare the strain on the rappeller without a swiss seat (left) with the one with a swiss seat (right). The strain on the lower back is much less when a swiss seat is used.

In an effort to eliminate these problems, some departments use the swiss seat in conjunction with the lifebelt. The effect of the swiss seat is two-fold: By using one-inch (25 mm) nylon strapping and tying it into a swiss seat through the ring of the lifebelt, the condition of the lifebelt is no longer critical, and the swiss seat transmits the load to the buttocks, directly below the natural center of gravity, resulting in a semisitting descent. Because of this semisitting position, the stresses of rappelling are applied to the body structure much like normal body loading. Because of these two advantages, the swiss seat significantly increases safety while rappelling. Nylon strapping is inexpensive and can be changed periodically. This eliminates the need to develop or institute elaborate testing procedures. In addition, 12 to 16 feet (about 4 m) of strapping can be rolled and easily carried in the pocket of a firefighter's coat.

Tying the Swiss Seat

Tying the swiss seat is relatively simple. There are several acceptable methods, one of which is given below.

Step 1: Divide a 12- to 16-foot (about 4 m) length of one-inch (25 mm) nylon strapping in half and place the strapping around the waist, with the center of the strapping at the small of the back (Figure 6.71).

Step 2: Pass the running ends of the strapping through the ring of the lifebelt and cross them (Figure 6.72).

Figure 6.71 Fold the strapping in half around the waist.

Figure 6.72 Pass the running ends through the ring. The load will be on the strapping instead of on the lifebelt.

Step 3: Pass the running ends of the straps between the legs, front to rear, and pull the straps snugly around the outside of the legs (Figure 6.73).

Step 4: Pass the running ends of the strapping under the strapping that is between the legs and pull snug toward the outside (Figure 6.74).

Step 5: Take the left running end behind the back to the outside of the right hip, take the right running end directly to the outside of the right hip, and begin to tie a square knot (Figure 6.75).

Step 6: Take as much of the slack out of the swiss seat as is possible without undue discomfort, working the slack from the lifebelt ring toward the knot started on the right hip. When the slack has been adjusted to the knot, finish tying the square knot on the right hip and make at least one half hitch safety with each running end of the strapping (Figure 6.76). The loose ends should be tucked away to prevent accidental tangling in the rappelling line.

Step 7: Hook onto the rappelling line (Figure 6.77).

Figure 6.73 Work out all the strapping slack as the seat is tied.

Figure 6.74 Pull the strapping through and to the outside. The weight will be spread over a larger area of the thigh and uncomfortable constriction when weight is applied to the seat will be prevented.

Figure 6.75 The final square knot is very important: it secures the swiss seat.

Figure 6.76 Secure the loose ends with a safety for the square knot. Tuck the ends in to keep them from tangling in the lifeline.

Figure 6.77 Wrap the lifeline in the hook as for normal rappelling.

Use of the Figure Eight and Carabiners for Rappelling

Figure 6.78 Carabiners are used extensively in rappelling. A locking carabiner like the one on the left is highly recommended.

In addition to the features of the lifebelt already mentioned, there is another characteristic of the lifebelt that makes it potentially hazardous for rappelling: the wraps of the rappelling line around the hook must be in a specific direction or the rappelling line may invert and slip out of the hook gate. Newer versions of the lifebelt have locking gates that eliminate this possibility. Some departments, however, have replaced lifebelts with standard rappelling equipment used by military and mountain-climbing groups. The two most used items are the carabiner, a D-ring device made of high-tensile-strength metal with a gate (Figure 6.78), and the figure eight, a forged assembly consisting of two rings made of high-tensile-strength metal. The figure eight is used for attachment to the rappelling line and the carabiner is used to attach the figure eight to the swiss seat. These devices both are small enough to be carried in a turnout coat pocket and they are inexpensive enough to be distributed as personal equipment.

The attachment of the figure eight to the rappelling line is simple compared to wrapping the rappelling line through the lifebelt hook. A bight in the rappelling line is pulled through the large ring of the figure eight and placed around the small ring (Figure 6.79). The small ring is then hooked into the carabiner and attached to the swiss seat. If additional friction is necessary, the rappelling line, after being passed around the small ring the first time, can be passed back through the large ring and back around the small ring (Figure 6.80).

Figure 6.79 The preferred figure eight has "ears" that prevent the rope from riding up and forming a clove hitch or larkspur that stops descent. The ears also add some friction and make "locking off" easier if a person has to stay suspended in one place to do some work.

Figure 6.80 With the more common figure eight without "ears," a double rap is usually necessary to add friction when rescuing a victim.

One may use the hook of the lifebelt instead of the carabiner. A person may also tie the swiss seat through the small ring of the figure eight by attaching the rappelling line to the figure eight before tying the swiss seat.

The use of the carabiner and figure eight with the swiss seat adds a large margin of safety to rappelling. It is nearly impossible to tie the figure eight wrong and have the rappelling line become free from the figure eight.

There are other more desirable procedures that use rappelling techniques instead of lifebelts for mountain and cave rescue. Departments with these needs should contact local mountaineering or cave rescue organizations for additional assistance.

CONTROLLING DESCENT SPEED FROM ABOVE

The friction of the rope wrapped around the hook controls the descent of a lifebelt wearer. A small amount of tension can be applied to these wraps on the running side of the hook, increasing the friction sufficiently to slow or stop the person sliding down a lifeline. Usually the person and the belt slide down while the lifeline remains stationary. The person sliding down the lifeline controls with one hand the tension of the wraps on the hook to control the speed of descent.

An alternative method is to secure the lifebelt to something solid, making the lifebelt and hook stationary at the top of the intended descent location. A firefighter stays at the top of the descent location to control the tension on the lifeline, controlling the descender's speed. A step-by-step explanation of this method follows:

Step 1: Secure the lifeline to a solid object at the top of the descent location (Figure 6.81).

Step 2: The hook of the belt is secured to the lifeline by any one of several ways; for example, a bowline or the quicker method given in Steps 3 and 4.

Figure 6.81 Use a bowline to tie the lifeline to a solid object.

Step 3: Form a bight in the rope where the hook is to be secured, then pass it through the ring of the hook and over the hook (Figure 6.82). The gate of the hook should be up.

Step 4: Use a bight in the running part to tie a safety of some kind on the standing part as shown in Figure 6.83.

Step 5: Secure with an appropriate knot, such as a bowline, or a second lifeline, the running end of the rope to the person who is going to descend.

Step 6: Adjust all the slack out of the lifeline between the hook and the person who will be descending.

Step 7: Wrap the rope through the hook as is usually done for sliding down a lifeline (Figure 6.84) and arrange the slack so it will play through the hook without tangling.

Step 8: The person who is to descend backs to the edge, pulling the lifeline through the hook while the descent controller controls the tension (Figure 6.85).

Step 9: While the descent controller watches from the edge and controls the descent speed, the descender continues backing over the edge and is lowered to the ground (Figure 6.86).

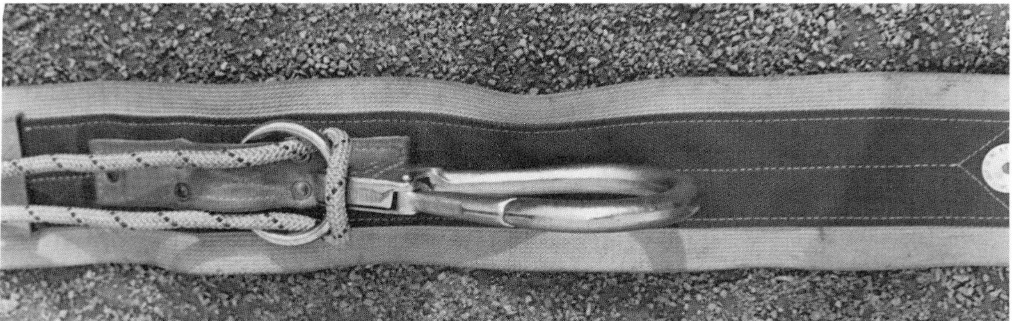

Figure 6.82 A bight is formed in the running part of the lifeline, then slipped through the ring and over the hook.

Figure 6.83 Another bight in the running part of the lifeline is used to tie a safety.

Figure 6.84 The hook is wrapped with the running part as if for regular rappelling.

Figure 6.85 The descent controller keeps tension on the line as the descender backs to the edge of the roof.

Figure 6.86 The descent controller keeps watch on the descender's progress.

Descent Control Devices

Lifelines used with descent control devices have certain advantages. They can be employed by victims or personnel with only a little training. They are lightweight, compact, versatile, and relatively inexpensive.

The descent control device uses rope friction to control the rate of descent. The device basically consists of an aluminum rod around or through which the rope is wrapped to create friction. The rope is held in position by the molded guides on the aluminum rod or by a metal cover (Figure 6.87). The number of turns taken around the rod or through the guides will control the amount of weight that can be lowered and the descent rate.

Descent control devices have some advantages

The devices have major rescue advantages in that the rate of descent can be controlled by the person descending, by a person on the ground, or by a person above the individual descending. In addition, the descent can be stopped and the device locked to let a person use both hands while suspended.

These devices can also be used to lower stretchers or equipment. Harnesses are available from the manufacturers of these devices.

Figure 6.87 One type of descent control device. *Courtesy of Montgomery County. Md., Fire/ Rescue Services.*

ELEVATORS

Elevators may be the safest form of public transportation, but occasionally power failures, equipment malfunctions, and fire emergencies cause problems for stalled passengers. Fortunately a runaway falling elevator is not a problem for rescuers because the equipment is designed to protect the passengers by preventing movement if the proper conditions for running are not satisfied. A stalled elevator can be a nuisance for the trapped passengers, or it can cause a critical life situation if fire is involved.

An equipment or power failure can stall elevator cars anywhere in the hoistway, but usually they will stall close to an opening. During the great Northeast power failure of November 1965, more than 80 percent of the cars in other than express hoistways stopped close enough to the floor that passengers could be removed by opening the car and hoistway doors.

Get advice from mechanic

In any elevator rescue the fire department rescue team can get its best advice from a qualified elevator mechanic. Immediately notify the mechanics, if available, of the shutdown. They will know the particular equipment involved, and can discover the cause of the failure and whether the trouble can be remedied quickly. Although the fire department can be trained to handle the situation alone, the most efficient methods for different elevators can best be determined when the rescue crew works with an elevator mechanic. Rescuers must be sure to turn off the power to the elevator during rescue operations. Fire personnel should control the power supply switch until the emergency is over.

Assure passengers

Regardless of whether there will be any delay in freeing the passengers, use the communications system installed in most elevators today to assure the passengers that they are safe, that there is no need for alarm, and that work to free them is beginning. The system may be an intercom or telephone. Intercom controls are usually at the main elevator control panel, but telephone hookups may differ. Some of them are controlled by guard stations or manager's offices or even a distant central protective agency office. The latter would present problems. If no communications system is available, one of the rescuers should speak to the passengers through the nearest hoistway door after the car's position is determined.

Talk with them as soon as possible to avoid the tension, anxiety, and even panic that may otherwise develop. Do not underestimate the unpleasantness of being trapped, possibly in total darkness, with a group of strangers and with time to imagine the worst for lack of better information. Injured or hysterical passengers require immediate assistance. Talking to them may not be enough. Sometimes the first action to help these people is to get a firefighter *inside* the car before getting the passengers out.

Tell the passengers briefly about the elevator's safety features. Most fear the car will break loose and crash to the pit below, but tell them the unit has many features designed to prevent it. If the rescue may take 30 minutes or more, tell them to sit and be more comfortable.

Access possible from sides or top

Removable side exits are held in place by at least four fasteners operable from both inside and outside the car. The hinged panels feature a key-lock arrangement on the inside to which elevator mechanics or building maintenance personnel will have the key. Access from the outside is by a permanent handle. Rescuers in the adjacent car will have to get the key of a hinged exit door so they can get to the outside handle of the stalled car.

Even though adjacent cars in the hoistway are close, rescuers will have to lay a short plank from one exit door to the other and help the passengers across. They may be unsettled by the seemingly precarious plank, but close assistance and reassurances by rescuers will help calm them. Tie victims and rescuers together with a lifeline during this operation.

Although complete power failures rarely occur today, a power feeder breakdown may prevent all the cars from running so that a side-by-side rescue would be impossible. If the position of the stalled car also makes opening the elevator doors impossible, then a top exit rescue may be necessary. Top exit panels are held in place by thumb screws on the inside. Outside there is a screwdriver slot for rescuers. Some recently installed units have only outside screws. In any event, rescuers will have to be outside and on top of the car to assist the passengers.

Elevator doors can be forced

First, open the hoistway door above the stalled car. Unless the doors have an emergency unlocking device operated by a special key to open the door from the landing side, the doors may have to be forced with an axe or other entry tool.

If a hoistway door emergency key is available, it will probably be a lunar key. The most common type is semicircular (moon shaped); others are T-shaped and some are circular (dropkeys). Semicircular keys are pushed in a few inches and pulled downward. T-shaped keys work by merely pushing them in; dropkeys rotate. None of them is standard, so only the right key will work. Learning their whereabouts should be a part of the planning for each building with elevators.

Try not to damage door

Forcible entry tools, if used, should be applied as close to the interlock as possible to keep damage to a minimum. The exact position of the interlock may be hard to determine, but it is usually on the header of the entrance frame. Single-slide doors do not indicate the lock side positively because the direction of travel is often unknown. Firefighters may be able to see the striker jamb if they look between the entrance frame and the door panel. They can use a flat object to probe tighter doors. Single-slide doors often close toward the hall buttons. Center-opening doors have the interlock near the top where the two doors meet.

On doors with two or more door panels, the cable may yield before the interlock. Then the panel on that cable will be freed and can be pushed open. Firefighters can reach behind the locked panel and manually trip the interlock. If firefighters are forced to break something, break the cable instead of the interlock. The cable is easier to repair. Single-slide doors have 20 cables.

A hydraulic portable power tool can be used to break the cables of a two-speed door by expanding the tool between the entrance frame and the outer panel. The increasing stress works through the cable to the interlock, and the cable should go first. Even if it does not, this method does not deface the door panel.

Talk with local elevator mechanics to learn about less expensive forcing techniques. Good public relations result if the rescue is made without heavily damaging the equipment. Some mechanics recommend sawing a hole in the hoistway door and reaching through to trip the interlock if forcing is absolutely necessary. Repairing or replacing a door is easier and less expensive than replacing a sprung door and door jamb. Warn passengers before sawing begins.

The top exit cover is either hinged or held by a chain. Open the cover outward and lower the rope or chain ladder, if one is provided in a metal box on the top of the car. If no preconnected ladder is there, use a 14-foot (4 m) fire department ladder. Passengers will need help to climb and maneuver on top of the car. The top area should be unobstructed, but it may be necessary to step over the car crosshead or other equipment to reach the next landing. Another ladder will probably be needed, depending on the car's position. Help the most nervous passengers first.

FIRE ESCAPE STAIRS

Fire escapes are found alongside many older buildings, but are at best a poor substitute for standard interior or exterior stairs.

Some fire escapes extend to the ground, which is preferred, but the lower flight might swing up if permanent stairs might obstruct sidewalks or aid intruders.

Problems with fire escapes

There are problems with fire escapes. The landing area of swinging stair sections is sometimes blocked by autos or materials temporarily stored there. Persons afraid of heights are reluctant to use them. In northern climates snow and ice limit their use. Access to fire escapes should be through doors, but sometimes access is through windows, which can be blocked by screens. Fire escapes may be in disrepair or insecurely fastened to the building, and heavy equipment such as aerial ladders should not be rested on them. Fire victims have lost their lives when fire escapes have collapsed.

The stairs of a counterbalanced fire escape from the ground level to the first landing are elevated horizontally with a counterweight. The stairs are lowered by stepping onto the suspended stairs a short distance and overcoming the counterweight.

The firefighter can lower the stairs from the ground by pulling on the end with a pike pole (Figure 6.88). Then use the fire escape to reduce victims or extend hoselines to upper floors.

Figure 6.88 A counterweighted fire escape may be pulled down with a pike pole.

IMPROVISED FIRE ESCAPE

An improvised fire elevator can be assembled with a suitable ground ladder and length of rope. This device will allow several conscious, but trapped, victims to be removed from a building. But the technique requires enough rescue workers to properly raise the ladder to be used. The ladder should be long enough to extend at least one rung above the opening from which the rescue is to be made. After this assembly has been positioned, the same number of rescue personnel as raised the ladder are needed to manipulate the device and safeguard the victims. The steps to perform this operation are listed on the opposite page.

Step 1: Select an appropriate ground ladder and rope. Position the ladder on the ground the correct distance from the building for a flat raise with the fly on the bottom. The heel should be toward the building with the ladder perpendicular to the wall and in line with the location for the elevator. The fly should be extended so four rungs of the bed ladder are exposed.

Step 2: Tilt the ladder up on either beam about 45 degrees. Place the rope at the heel in a coil that will pay out easily. Take the running end to the tip, and push it up from the underside between the first and second rungs from the tip. Take the running end back to the heel and push it through the ladder between the third and fourth rungs from the heel, and back up between the first and second rungs. Then pull it toward the tip several feet.

Step 3: The remaining coiled rope is then payed out to avoid any binding. The standing end of the rope is then tied to the running end with a sheet bend. To avoid overloading the rope or ladder tie overhand knots to form loops about ten inches (25 cm) long at the intervals listed below:

Second floor rescue: 10 feet (3 m) apart
Third floor rescue: 15 feet (4.5 m) apart
Fourth floor rescue: 20 feet (6 m) apart

Step 4: Raise the ladder to vertical. Extend the fly enough to reach above the opening. Adjust the rope and place the ladder against the building normally.

Step 5: Direct the victims to step into a loop with one foot and hold to the rope above their heads with their hands. One firefighter should brace the ladder and control the rope speed. Another firefighter should feed additional rope and help keep victims clear of the wall. If available, another firefighter should heel the ladder from underneath.

This technique can be easily modified to lower an unconscious victim. The sheet bend is adjusted to the top of the ladder after it is in position. Firefighters at the top pull the knot in and untie it. The end feeding over the top rung is retained and the opposite end is secured momentarily to a rung. Be sure not to confuse the ends. The retained end is attached to the victim with a bowline on a half hitch. The end set aside is then attached as a guide, and the firefighters on the ground pull the slack from the rope. The victim is placed on the window ledge and when ground personnel are prepared the victim is gently let out the window and lowered to the ground.

Water and Ice Rescue

Chapter 7

Chapter 7
Water And Ice Rescue

Rescuing persons in water accidents presents special problems for the firefighter. Water rescue often appears to be less dangerous, but the firefighter should remember that water can be as deadly as fire (Figures 7.1 and 7.2).

Figure 7.1 Calm streams can quickly become torrents. *Courtesy of Stillwater, Okla., News-Press.*

Figure 7.2 Floods can create more life hazards than is commonly supposed. Note here the building collapse, downed electrical wires, and broken gas line at meter. *Courtesy of Montgomery Co., Md., Fire/Rescue Services.*

Water rescue involves more than rescuing a drowning person. It also involves ice rescue, diving injury rescue, traumatic injuries in water, water vehicle rescue, and victim body recovery (Figure 7.3). The number of victims may range from one to several hundred (Figure 7.4).

Figure 7.3 Pleasure-boating accidents account for many water rescue calls. *Courtesy of Montgomery Co., Md., Fire/Rescue Services.*

Figure 7.4 The fire department could be called upon to assist in water rescues involving many victims. *Courtesy of U.S. Coast Guard.*

Drowning is technically death by suffocation (absence of blood oxygen). Therefore, the primary objective of water rescue is to restore normal breathing. Attempts to stabilize or restore breathing must be initiated immediately. Merely taking the victim to land will not in itself save the victim. While restoring breathing is the primary step in water rescue, other life threatening problems may well accompany or be the result of the incident. Therefore, the victim must receive proper medical attention and transportation if recovery is to be insured. (For complete information on first aid considerations see IFSTA's **FIRE SERVICE FIRST AID PRACTICES**.)

The firefighter may be confronted with a water rescue under many conditions, which can be generalized into three basic types.

- The victim is not in immediate danger of drowning, but the firefighter is required to use special skills to get the victim out of the water safely.

- The victim is struggling to keep from drowning, or the victim has gone under but there is still hope for resuscitation.
- Victim recovery is probably too late for resuscitation.

Situations where a victim may not be in immediate danger of drowning but will require the special rescue skills of the firefighter are.

- Rescue from partially submerged and submerging vehicles
- Rescue from ice
- Rescues involving traumatic injury to swimmers
- Rescue in self-contained underwater breathing apparatus (Scuba) diving accidents

SUBMERGED VEHICLES

Partially submerged vehicle and submerged vehicle rescue combines the problems of automobile accidents and water rescue. Several considerations must be made before rescue operations can begin: the location and the physical condition of the victim, the distance from the vehicle to a possible rescue point, whether the vehicle is resting on the bottom, the water current, and the water temperature.

Most automobiles will float briefly when first entering the water, so the fact that a vehicle is visible on the surface when rescue workers arrive is no assurance it will remain so.

Figure 7.5 A motor vehicle will sink engine end first. *Courtesy of Montgomery Co., Md., Fire/Rescue Services.*

Studies on submerging vehicles indicate a vehicle will sink engine end first (Figure 7.5). However, the studies also show vehicles generally come to rest level. A listing vehicle is probably sinking. Often the vehicle will be only partially submerged.

The first company to arrive should request aerial equipment if it was not dispatched initially. Firefighters should not enter the water unless the vehicle is submerging or submerged and the passengers unable to get out. Only experienced swimmers with lifelines and life preservers should be allowed to enter the water to attempt rescue. If the victims have been able to exit the vehicle or can remain in the passenger compartment without danger, they should be urged to stay with the vehicle unless it is submerging rapidly. If the victims are out of the vehicle, tell them to sit on the roof or hang onto the vehicle. They will be safer there than trying to swim to shore.

If there is no immediate need to rescue victims from drowning and the vehicle is not submerged, firefighters should stabilize the vehicle before rescue. The vehicle can best be stabilized by lashing it to several objects on land upstream from the vehicle. If the vehicle is in a narrow body of water, tie it to both shores. This way the vehicle can be positioned to allow the optimum stabilizing angle.

Stabilizing is attempted only to prevent the vehicle from shifting or being washed away with the current; it can also slow the rate at which a vehicle is submerging or stop it from submerging. When the vehicle is floating, firefighters can try to tow the vehicle to shore, but make no attempt to move a vehicle that has come to rest on the bottom. Movement may cause the vehicle to sink further or dislodge the victims from the vehicle.

Aerial apparatus could be used

After the vehicle is stabilized, develop a way to remove the victim. Aerial equipment often provides the best way. If the vehicle is close to the shore, firefighters can use a ladder bridge. The procedures for ladder bridge techniques can be found in IFSTA's **GROUND LADDER PRACTICES.** If aerial equipment is used, place the tip of the ladder or platform at the vehicle as a platform from which rescuers can work.

If a vehicle has submerged, the firefighter must note the location. Unless the vehicle is in very shallow water, unequipped or inexperienced personnel should not dive to it. Firefighters' breathing apparatus should not be used under water because the regulators will malfunction. Use a qualified diver for rescue. Remember that submerged vehicles will trap some air in the passenger compartment, so rescues can be effective after the vehicle has submerged. Trying to tow a completely submerged vehicle to shallow water can prove effective if undertaken soon after the car has submerged. The vehicle will remain somewhat buoyant because of air trapped inside.

Special Response Unit

The best way to insure rescue from submerged vehicles is by adequate training and proper planning before the accident occurs. If the submerged vehicle rescue occurs in a department's jurisdiction, a special unit should be ready for dispatch from within the emergency network. This unit must have at least two certified divers on call. For large departments with substantial resources, this unit conceivably can come from within. Smaller departments with fewer resources should consider developing such a unit with other branches of emergency services or a local scuba club.

Have system for diver response

A communication system with the divers must be developed for emergency use. Ideally, on-duty firefighters could be trained to run the system within the normal fire department communications framework. If outside divers are used, a radio alert system is essential for a quick response.

The fire department usually keeps the rescue equipment because the department's emphasis on emergency preparedness will help keep the unit ready. The time for setup and organization at the emergency scene will be short. Divers can keep their personal equipment with them, but the unit should have at least one extra set of diving gear. Departments that must cover considerable water frontage can develop a water rescue squad equipped to handle most water emergencies from drowning to sinking boats.

Most fire departments will not have the resources to equip a separate vehicle for underwater rescue. Therefore, underwater rescue equipment will typically be carried on the department's primary rescue vehicle. Most of the rescue equipment used for ordinary rescue situations will also serve for underwater rescue. The equipment needed specifically for underwater rescue should be separated from the other in a special compartment or locker for underwater rescue equipment. Departments should have the following on hand.

- Lifelines
- Yellow diver marker buoys
- Red diver marker buoy
- Underwater handlights
- Flotation vests
- Set of additional diving gear
- Spare air bottles
- Backboard with flotation device
- Flotation rings for Stokes baskets
- Body bags
- Underwater cutting torch
- Rope ladder

Rescuing Oneself From a Submerged Vehicle

Firefighters themselves may some day be trapped in a submerging vehicle, joining an estimated 400 U.S. citizens every year. Two recently completed studies, one by the Dutch government and another by the American Red Cross, reveal similar facts about submerging vehicles.

Submerged vehicle characteristics

- Adequate air will generally be trapped under the roof of the vehicle for any occupant to survive for up to fifteen minutes, even with the windows open.
- The primary factors in survival are avoiding injury during initial impact and staying calm.
- Most cars will come to rest on the bottom on their wheels, unless they enter the water on their top with the windows open.
- The pressure of the incoming water will generally keep an occupant from escaping until the inside is full.

- The occupant will generally not be able to open a door until the passenger compartment is full. However, windows will roll up or down.
- When a vehicle submerges in water 12 feet deep (3.6 m) or less, the vehicle will come to rest on the bottom within ten seconds after it disappears under the surface.

People obviously can survive if trapped in a submerged vehicle, but they must stay calm. The Dutch suggest that if drivers see they are going to end up in the water, they should speed up to increase chances of landing upright. The best way to escape is through the window before the vehicle sinks below window level.

Be calm and plan escape

If escape through the window is impossible, roll up the windows and get in the back of the car. Most cars will submerge engine first. If the victims cannot escape before the vehicle submerges, they should try to orient themselves to the location of land. Any objects that float as the car sinks may serve as floats on the surface. Since doors will be hard to open until water fills the passenger compartment, the extra time should be used to calm down and to plan an escape. Take off shoes and extra clothing. Underwear is all that should be worn to the surface.

When the water level is near the top of the doors, make last-minute plans. Try to remember which way land will be and grab any buoyant objects floating in the car to take to the surface. Push open the door and get ready to escape. Take several deep breaths at a slightly accelerated rate instead of a final deep breath. Then victims should push under and out, keeping their legs coiled and ready to push off against the car as soon as they get outside. The eyes should be kept open.

Swimming to the surface will take slightly longer than the trip down. If the water is more than ten feet (3 m) deep, victims will have to exhale part of their air on the way up because of the pressure difference. When on the surface, victims should relax and catch their breath before heading for land.

These techniques will work regardless of the position in which the car comes to rest. If the car is on its side or top, the length of time before exiting will be shorter. If water is rushing in, brace against it and wait until it equalizes in the car. If a door cannot be found that will open, roll down the windows or break them out with a pointed object. If no object is available, push the windows out with the legs.

ICE RESCUE

The ability of people to survive in cold water is limited. The loss of body heat in water is about 25 times greater than the loss of body heat in the normal atmosphere, so once people are suddenly submersed in icy water, speedy action is necessary to prevent death. The immediate goal is to keep victims from sinking and drowning. Victims can rescue themselves, but panic often makes this unlikely.

Rescuers should never attempt an ice rescue without adequate preparation. To do so would turn a relatively simple rescue operation into a complex one, which may imperil the rescuers. Rescuers should fasten a rope around their waists whenever they are on the ice. Their safety depends on being able to recognize unsafe conditions. With thorough training and quick application of proven procedures, they can avoid fatal results. Firefighters must remember that ice rescue can be dangerous and not take any chances. Only the minimum number of rescuers should be on the ice at any time.

The basic principle is to spread the weight of anyone on the ice over as large an area as possible. Ladders placed on the ice will help spread the weight well (Figures 7.6 - 7.12). If the ice is cracked or seems thin, use other methods to go onto the ice (Figures 7.13 - 7.17).

Figure 7.6 Tie a rope to one end of the ladder.

Figure 7.7 Slide the ladder to the victim.

Figure 7.10 One person using a ladder to rescue a victim. *Courtesy of Cranston, Rhode Island Fire Department.*

Figure 7.8 and 7.9 The rescuer on shore holds the rope while the other rescuers steady and guide the ladder.

Figure 7.11 and 7.12 Almost any maneuverable object that helps to distribute weight may be used when rescuing a person from the ice. *Courtesy of Cranston, Rhode Island Fire Department.*

Figure 7.13 A toboggan can distribute the rescuer's weight. *Courtesy of Cranston, Rhode Island Fire Department.*

Figures 7.14 and 7.15 When ice is thin, rescuers should try to reach victim from a distance.

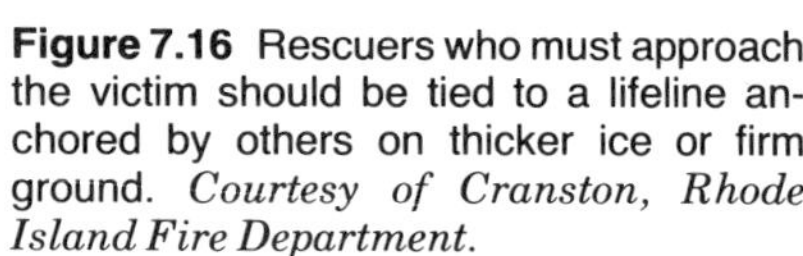

Figure 7.16 Rescuers who must approach the victim should be tied to a lifeline anchored by others on thicker ice or firm ground. *Courtesy of Cranston, Rhode Island Fire Department.*

Figure 7.17 A lifeline should also be used when a rescuer must swim to the victim. *Courtesy of Cranston, Rhode Island Fire Department.*

Every effort should be made to reassure victims and to keep them from wasting strength by thrashing about. They should try to crawl forward on their stomach until their hips are at the edge of the ice. Then, with their arms above their head, they should quickly roll away from the edge of the ice (Figures 7.18 - 7.20).

Some communities have developed ice rescue squads that use special insulated flotation suits. One firefighter dons the flotation suit and goes onto the ice with a life jacket and a lifeline. The rescuer tries to stay on the ice but can go into the water without hesitation if needed. Victims are often small children and the rescuer will not be able to reach them without breaking through

Figures 7.18 - 7.20 The first step in self-rescue is to get the trunk of the body out of the water and onto the ice. The method shown in these illustrations works well on thick ice. When the ice is thinner one should leave the hole on one's back, then work oneself to thicker ice before turning onto the stomach and using the arms to lever oneself up.

the ice because of the extra weight. When this happens, the rescuer will have to swim, preferably by backstroke.

On reaching the victim, the rescuer stabilizes the victim by grasping the victim or attaching the life jacket. Keep the victim afloat, because a rescuer in a flotation suit cannot dive. Victims should be reassured and asked to remain calm to conserve body energy. Pull them in with the lifeline if possible. If not, support them until a raft, boat, or other rescue device can be deployed (Figures 7.21 and 7.22).

Figures 7.21 and 7.22 A commercial water and ice rescue catamaran. *Courtesy of Chicago, Ill., Fire Dept.*

While rescuing victims remember that they will be suffering from hypothermia—the lack of body warmth. Though they may be safe from drowning, serious complications can develop because of this condition. Victims also will be unable to grasp or pull effectively because of the cold. The rescuer will have to do most of the work.

Prudent departments now dispatch more than one ambulance to ice rescues because rescuers often will need transportation, too. Ambulance attendants with electric blankets should turn them on when first dispatched.

MAMMALIAN DIVING REFLEX

Several people who have drowned in cold water have been revived after a long time lapse, more than 30 minutes in one case, because of a condition known as the mammalian diving reflex. Marine mammals have developed the ability to submerge for extended periods of time without brain or tissue damage, and this same phenomenon may be aiding people. The phenomenon is attributed to the body's ability to channel blood away from muscle, arms, and skin to circulate it only between the brain and lungs. The skin, muscles, and inessential organs can survive oxygen loss for about an hour. During this time all the oxygen in the blood is transferred to the brain, which can survive only three or four minutes without oxygen.

Studies show that the reflex is predominately found in water less than 70° F (21°C). The colder the water, the more pronounced the reflex and the greater the possibility of survival. Younger victims are also more likely to survive. The reflex is more pronounced in victims under 3½ years old.

Persons thought dead may be revived

Persons who have recovered from drowning after experiencing this reflex have almost without exception exhibited death-like symptoms—dilated pupils, no pulse or heartbeat, no detectable breathing. There is no way to tell if the reflex has been triggered.

But remember that cold-water in itself has adverse effects on the ability of a victim to survive. Studies of shipwrecks in cold water indicate that cold-water immersion ranks with drowning as a cause of death. How long people can be expected to survive in cold water is shown in the table below.

Survival Time	Water Temperature °F	°C
20 to 30 minutes	40	5
1½ to 2 hours	40-64	5-20
Indefinitely	64	20

Studies also indicate that rapid immersion in cold water can kill good swimmers in four to five minutes (Figures 7.23 and 7.24).

In conclusion, immersion in cold water has been known to increase the chances for survival of the victim to times beyond those expected for normal drownings. Remember, however, that cold water may in fact have an equal effect in premature drowning.

Figures 7.23 and 7.24

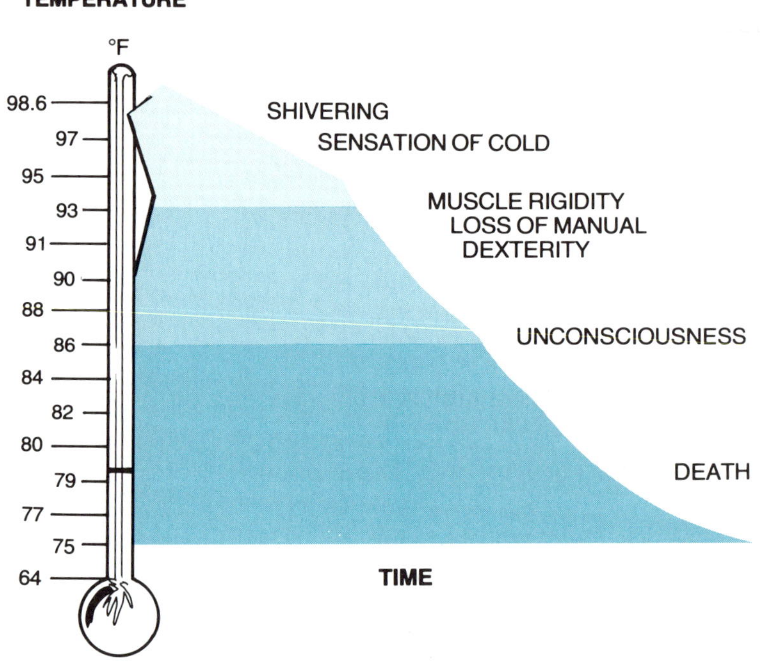

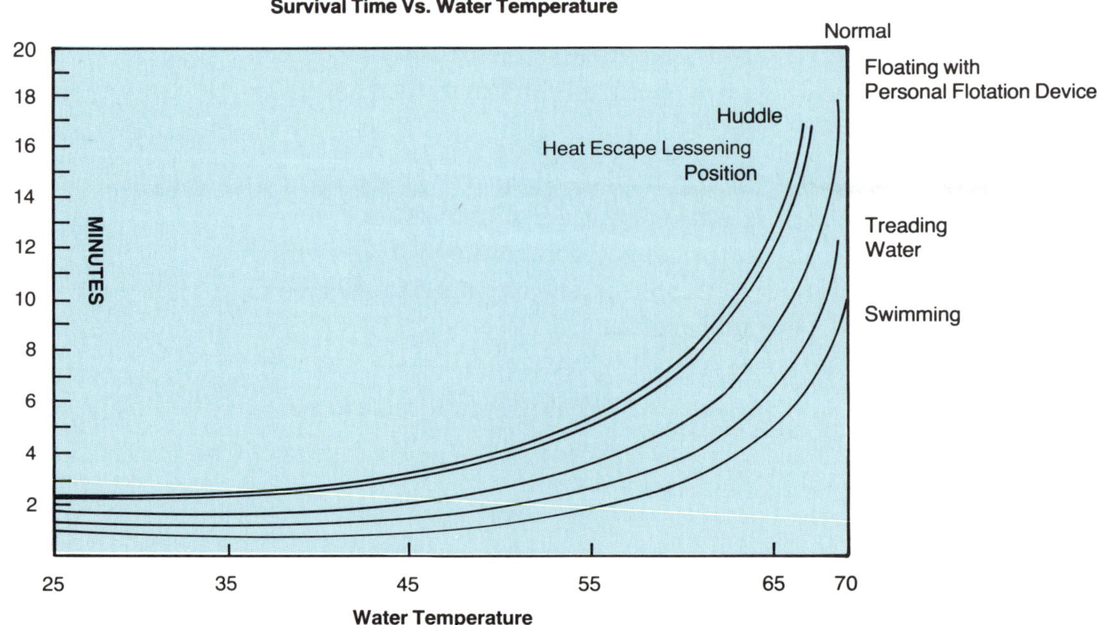

INJURIES TO SWIMMERS

Firefighters will often be summoned to help injured swimmers. The most common injuries come from diving into shallow water and involve neck and spinal damage. The rescuer must make every effort to stabilize the victim in the water before removal. Support the victim's body in the water to let the victim breathe easily. Encourage the victim to relax and let the body's natural buoyancy keep it level. Attach a cervical collar and place the victim on a full backboard while still in the water. Then remove the victim from the water (Figures 7.25 - 7.34). (See IFSTA's **Fire Service First Aid Practices** for details on using backboards.)

Figure 7.25 A person injured in a diving accident. *Courtesy of Montgomery Co., Md., Fire/Rescue Services.*

Figure 7.26 Grasp the victim by the upper arm; place hand on spine. *Courtesy of Montgomery Co., Md., Fire/Rescue Services.*

Figure 7.27 Turn the victim face up, keeping spine supported. *Courtesy of Montgomery Co., Md., Fire/Rescue Services.*

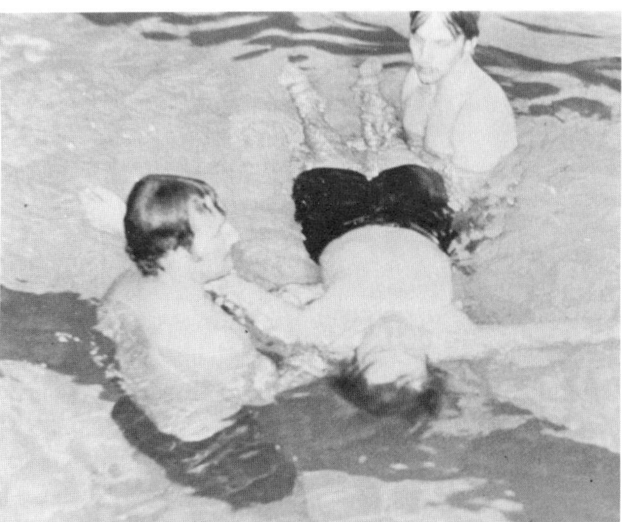

Figure 7.28 Keep victim's spinal column as straight as possible by supporting the back and legs. *Courtesy of Montgomery Co., Md., Fire/Rescue Services.*

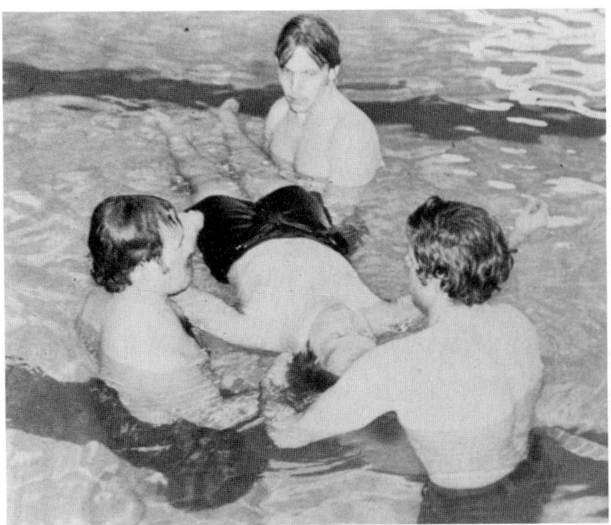

Figure 7.29 Attach a cervical collar. *Courtesy of Montgomery Co., Md., Fire/Rescue Services.*

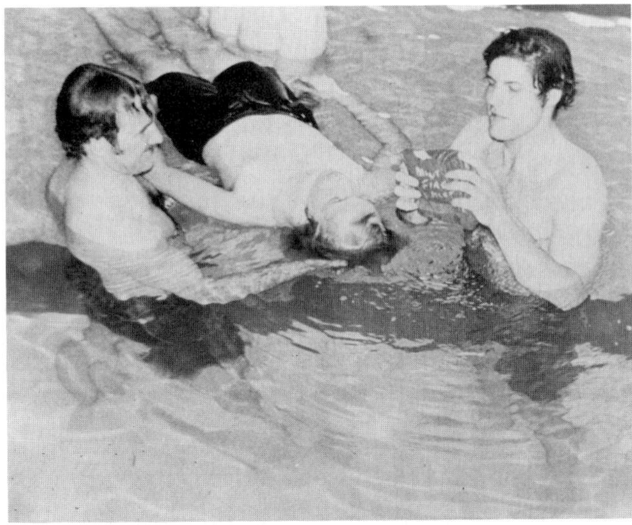

Figure 7.30 Place extra support under the neck. Here a special cushion is being used, but a folded beach towel or similar object will suffice. *Courtesy of Montgomery Co., Md., Fire/Rescue Services.*

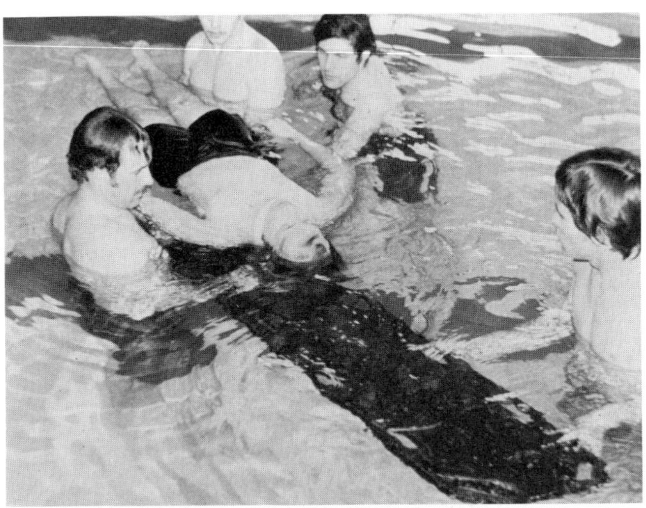

Figure 7.31 Slide backboard underneath victim. *Courtesy of Montgomery Co., Md., Fire/Rescue Services.*

Figure 7.32 Secure victim to backboard. *Courtesy of Montgomery Co., Md., Fire/Rescue Services.*

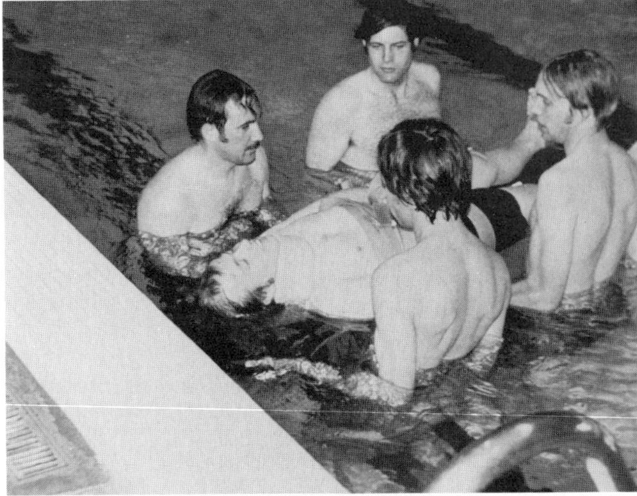

Figure 7.33 Float victim to edge of pool or to bank. *Courtesy of Montgomery Co., Md., Fire/Rescue Services.*

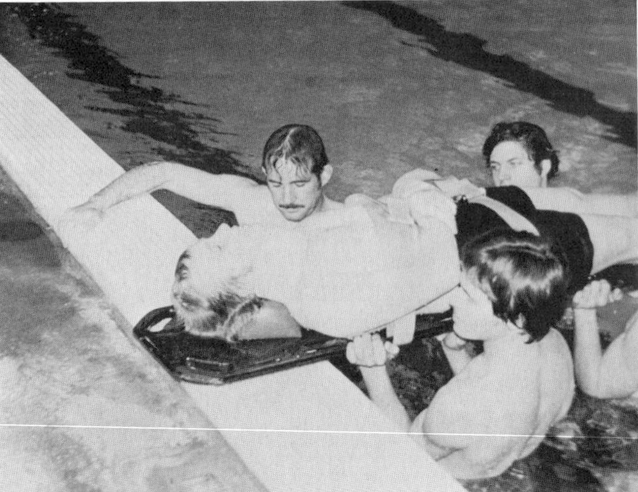

Figure 7.34 Keep backboard horizontal and slide it as straight as possible onto edge of pool or bank. *Courtesy of Montgomery Co., Md., Fire/Rescue Services.*

SAVING A DROWNING VICTIM

A drowning victim needs quick action from rescuers. Only trained firefighters should attempt a swimming rescue. The rule of thumb in drowning rescue is "Throw, tow, row, then go." (Figures 7.35 - 7.37).

First, try to throw a buoyant object to the victim. Life preservers or water rings are usually stationed in swimming areas.

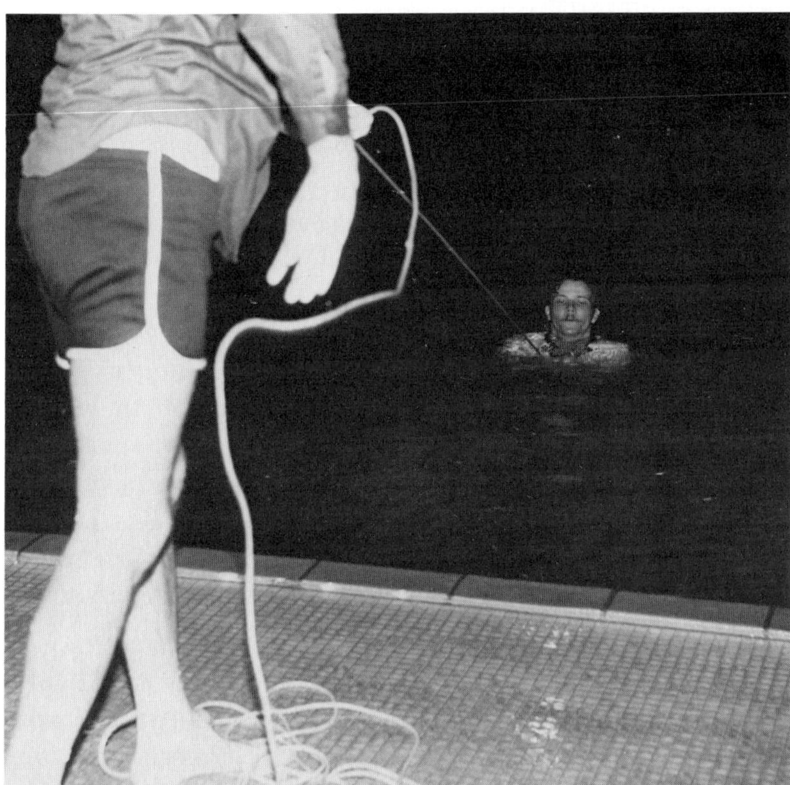

Figure 7.35 Throwing something to the victim and then dragging the victim to safety is best when possible. *Courtesy of Bill Maggi.*

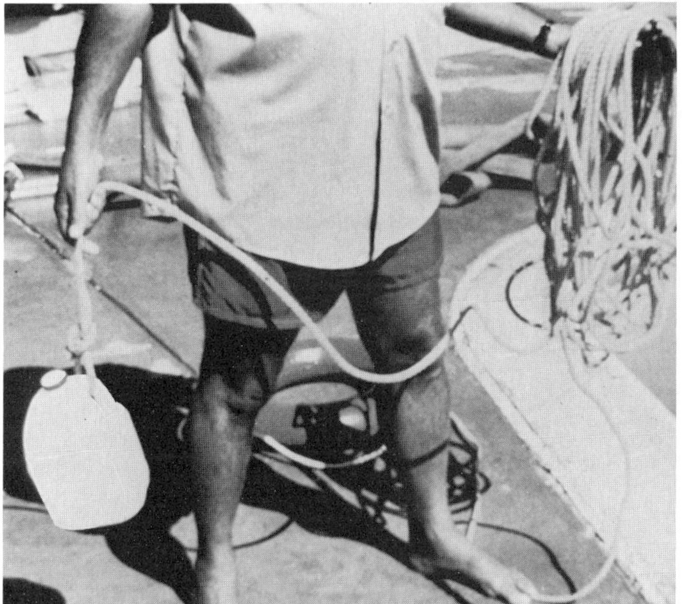

Figure 7.36 A plastic jug, about one-third full of water for throwing weight, makes a good rescue and flotation device. *Courtesy of Montgomery Co., Md., Fire/Rescue Services.*

Figure 7.37 A rope with a weighted end can be thrown to the victim. *Courtesy of Montgomery Co., Md., Fire/Rescue Services.*

Then try to reach the victim with a rope or line, using a rope throw. The rope will usually float on or near the surface long enough for a person to reach it and be towed to safety. Use polyethylene or polypropylene rope if available.

If these methods are not available, use a boat. This is not a particularly practical approach, because boats are seldom handy, and small ones capsize easily. But using a boat is better than trying a swimming rescue.

Swimming rescues should be tried only by a qualified person and as a last resort. Untrained swimmers often become victims of their own rescue attempt because panicky drowning persons will take the rescuers under with them.

Certain techniques improve chances that swimming rescues will succeed. Rescuers can push a wooden backboard, a short wooden ladder, or other buoyant object ahead of them to keep the victim away while being dragged to safety (Figure 7.38). If this technique is not possible, the swimmer should attempt to grasp the victim from behind. This may require going under the water and turning the victim or diving under the victim and approaching from the rear. If the victim panics, the rescuer may have to go under water or turn the victim face down in the water. Generally, a panicky victim will let go and try to return to the surface when the rescuer goes under the water. The rescuer should swim down and away from the victim before resurfacing.

If the person has just gone under, swimming may be the only way to save the victim. The rescuer should go to where the victim was last seen and search in a circular pattern. Unless there is a very strong current, the victim should be a distance no more than 1½ times the depth of the water from the point where last seen.

Figure 7.38 If the rescuer must swim to the victim, the rescuer should take some buoyant object for the victim to hold onto. This will prevent a panicky victim from grabbing the rescuer and drowning them both. *Courtesy of Bill Maggi.*

A rescue squad should treat all water accident victims as recoverable and, because of the mammalian dive reflex, should not hesitate to attempt rescue.

THE LADDER FLOAT DRAG

The ladder float drag is one method of water and ice rescue (Figure 7.39). An automobile inner tube or mounted spare tire secured to the end of a ladder will support a person, who can then be dragged to safety. Handlines can be secured to the end of the ladder to guide it to shore. Water rescue squads may want to have a way to inflate an inner tube for this purpose.

Figure 7.39 Method of securing ladder to tire to make ladder float. *Courtesy of Springfield, Mo., Fire Dept.*

Doughnut Technique

The only equipment required is a section of fire hose laid out in a straight line. One firefighter bends the female coupling back about one foot (30 cm) to kink the hose, and squeezes the kink tightly together with the hands to prevent air leakage. A second firefighter takes the male coupling, exhales forcibly into the hose until it is filled with air, and kinks the hose at the male end to prevent air leakage. The male and female couplings are then attached tightly together and the doughnut is ready to float.

Raft Technique

The raft technique uses the doughnut, but bundles the hose into a compact raft. This technique has advantages over the doughnut, is easier to maneuver in the water, will support more weight in the water, and tends to deflate more slowly than the doughnut.

Pull the doughnut into an oval so the two 25-foot (7-8 m) lengths are side by side, fold the oval in half, and then fold it in half again so there are eight six-foot (2 m) lines side by side. Lash the lengths together with a rope (Figure 7.40).

Figure 7.40 Raft made of inflated fire hose. *Courtesy of Bill Maggi.*

The rope lashing will resemble weaving. About 12 inches (30 cm) from one end of the raft, run the rope under the outside flake and then over the top of the next. Repeat the procedure through all flakes. Then run the rope back through the flakes in an opposite pattern and tie it at the loose ends to hold each section together without bunching. Make a similar lashing about 12 inches (30 cm) from the other end of the raft.

Straight Float

The same technique and some common pumper equipment can be used to form a straight float. The extra equipment needed is a double male adapter and two pump discharge caps. This method usually is restricted to hose with 2½-inch (65 mm) couplings. The firefighter can use it to reach a victim 50 feet (15 m) away (Figure 7.41).

Pull the hose from the pumper and attach the double male to the female coupling. Seal the coupling with a discharge cap to prevent air leakage. Inflate the hose and cap the second end with the other discharge cap. Then push the hose toward the victim. If the hose is pushed into a moving body of water, the forward end may have to be guided by ropes.

Figure 7.41 A variation of the straight float, giving the victim a loop for better support. *Courtesy of Montgomery Co., Md., Fire/Rescue Services.*

Rapid Inflation Techniques

Rescue attempts will be quicker if fire departments use compressed gas to inflate the hoses. Compressed gas is quicker than lung inflation, fills the hose completely, and lets firefighters couple and fill multiple sections of hose.

A CO_2 extinguisher is easiest to use. Cap or kink the female end of the hose as before and fit the horn of the extinguisher snugly against the male coupling. The firefighter holding the horn against the coupling must wear a good pair of gloves to prevent frostbite.

Breathing apparatus can also be used to inflate the hose (Figure 7.42). Remove the facepiece of the mask from the low-pressure tube (with a Scott this will require a screwdriver). Insert the tube into the male coupling of the hoseline. Put a seal around the tube. Open the regulator bypass valve completely and use the bottle valve to regulate the flow. Figure 7.43 shows another method that uses a special device to which the high-pressure hose is connected to fill the fire hose with air.

Figure 7.42 Using scuba cylinder to inflate a section of fire hose. *Courtesy of Dave Mette.*

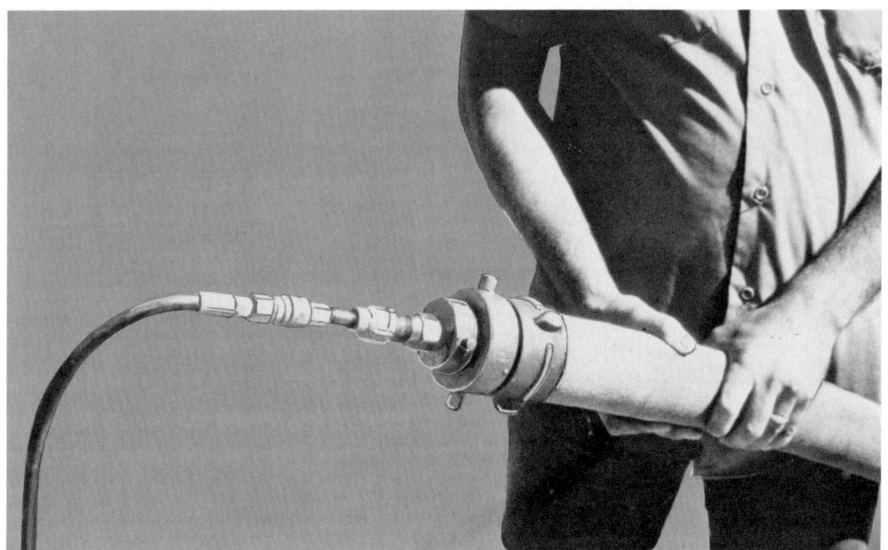

Figure 7.43 A shop-built device for rapid inflation of fire hose. *Courtesy of Montgomery Co., Md., Fire/Rescue Services.*

DROWNING MACHINES (See Appendix D pge 254)

Low-head dams create a hydraulic action known as the "drowning machine" when they overflow during high water (Figures 7.44 and 7.45). The rolling action of the water flowing over the dam causes a strong current to return to the point of downwash. An object can be pulled into the downwash, pushed to the bottom, pulled to the surface by the same current, and taken back into the downwash. This cycle can go on indefinitely. Persons caught in the cycle will be continually dunked until they are rescued, manage to break free, or drown. The current can affect objects as much as 50 yards (45 m) downstream from the dam.

Figure 7.44 A low-head dam at flood. *Courtesy of Montgomery Co., Md., Fire/Rescue Services.*

Low-Head Dam at Flood
The Drowning Machine

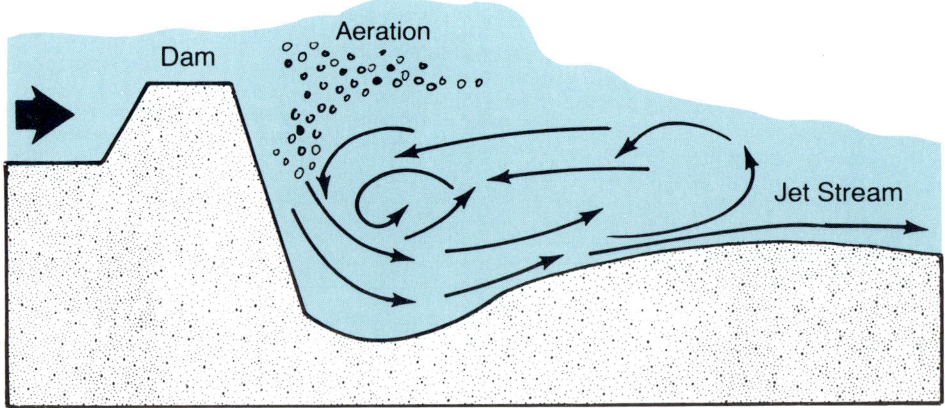

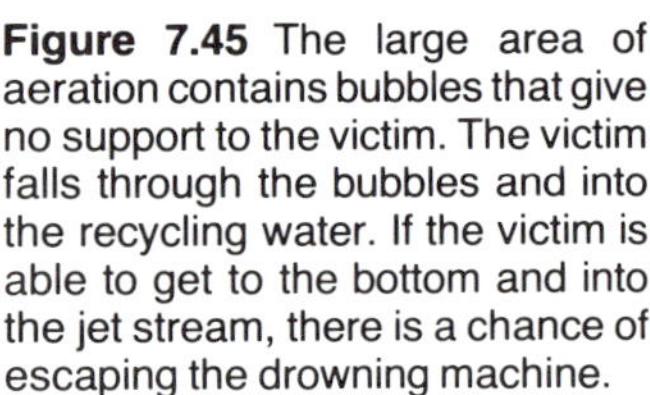

Figure 7.45 The large area of aeration contains bubbles that give no support to the victim. The victim falls through the bubbles and into the recycling water. If the victim is able to get to the bottom and into the jet stream, there is a chance of escaping the drowning machine.

UNDERWATER RECOVERY

When the victim cannot be located rapidly, body recovery becomes the priority for rescuers. This can often be a difficult task.

Planning

All agencies participating in a recovery operation must be coordinated and completely understand recovery plans to avoid confusion and increase efficiency. Some of these agencies are the lake authority, sheriff's office, highway patrol, local and state police, auxiliary police, and fire service rescue units. Care and exactness in planning an operation is necessary to obtain fast and satisfactory results. Some of the people with whom a rescue squad must deal are volunteer workers, sightseers, relatives, friends, and persons with self-appointed authority. Volunteer workers should be questioned to determine their capabilities and instructed and assigned accordingly. Sightseers should be told exactly what is expected of them and these expectations *must* be enforced. People who have assumed self-appointed authority should be dealt with quickly and finally. They should be admonished to restrain their actions or arrested for interfering with an officer.

Control spectators promptly

Take care to explain the situation to family and friends of the victim and assure them that everything possible is being done. This will insure less interference and promote better public relations between the department and the public. Clergymen can be a great help with family and friends. When arriving at a possible drowning, responsible personnel, preferably with interviewing experience, should question witnesses about the exact location and time of the incident.

Authority and Recovery Preparation

One person, usually called a "beachmaster," should be in charge of a recovery operation and should have the complete cooperation of all involved. A beachmaster's duties are wide and varied. A first concern is the safety of rescue personnel. The beachmaster should demand that all rescue personnel wear life jackets, regardless of whether they can swim well.

Rule of thumb for finding victim

The information gathered by responsible personnel should be submitted to the beachmaster and a decision made about the victim's possible location. A human body will usually be near the point where it was last seen. One good way to find a victim, illustrated in Figure 7.46, uses a rule of thumb to give a fair

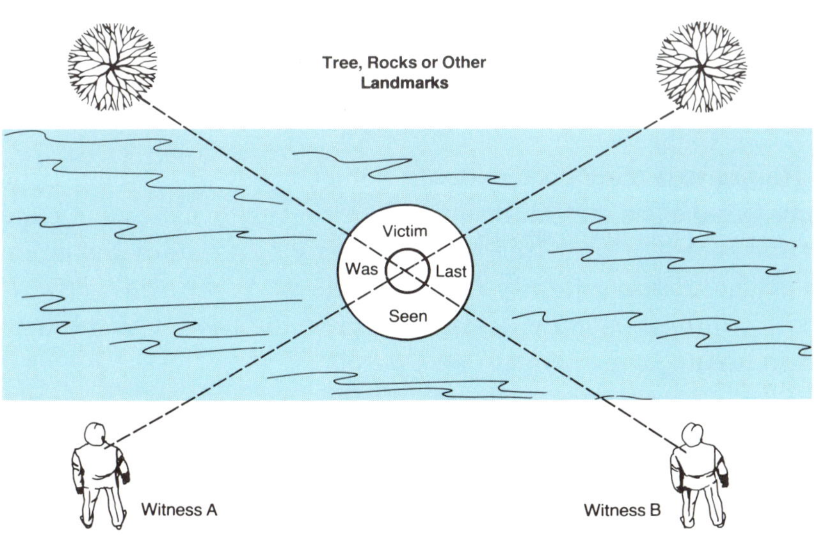

Figure 7.46 Triangulation of witness sightings can help locate victim.

estimate of where to search in unturbulent fresh water. The rule states that a sunken human body usually will be within a distance not more than 1½ times the depth of water from the point where it was last seen. For example, if the water is 20 feet (6 m) deep, a body should be no more than about 30 feet (9 m) from the designated point.

Human bodies that are obese or bodies of small children may not sink to the bottom, and, in some instances, they may stay afloat. A human body will rise to the surface when enough gas has formed within the intestinal tract to make the body buoyant. The temperature of the water and the content of a victim's stomach will affect the time needed to generate enough gas. During the summer the average rising time is from 18 to 24 hours. During the winter months, or when the water is deep and cold, the rising time will be longer. A human body will rise gradually in proportion to the amount of gas that has formed. Its movement may be affected by undercurrents. After reaching the surface, the body may also be affected by the wind. Current and wind can move human bodies several miles from the point where they were last seen. A victim who drowned in rapids will probably be found in the first deep hole downstream. In this situation, send a rescue detail downstream to search the eddies and center of the stream. The character of the bottom will be a deciding factor when planning how best to recover a sunken human body. Drags or grappling hooks will probably not work when bottoms have boulders, tree stumps, logs, and other obstacles. In these situations use trained divers.

The Search

The area to be searched should be marked off with marker buoys and efforts concentrated within this area. The search must be thorough, for it must cover the entire area of the bottom. Even after a thorough search it may be necessary to retrace the same pattern several times, especially if the victim is not clothed. The search may be done by scuba divers or with grappling hooks or drags. These two techniques should not be combined in the same area.

Remember that hooks should be used only after there is no chance to save the victim. Because of the nature of a drowning and the time required to search an area, many bystanders and members of the victim's family may be present. When the victim is found firefighters should be discreet in what they let bystanders see and hear.

Grab lines should be held by hand (Figure 7.47), not tied to boats, so squad members can feel and detect anything the hooks may contact. A human body can be hooked and turned over but not hooked securely enough for recovery. If motors are used on

Figure 7.47 Grab lines should be held in the hand, not tied to the boat. *Courtesy of Montgomery Co., Md., Fire/Rescue Services.*

rescue boats, very slow speeds are necessary to keep drags and grappling hooks on the bottom.

A wide area or span can be covered by tying three to five boats side by side. Remove the motors from all but the two on the outside. Use small motors. With a little practice the two motor operators can move and turn the entire unit accurately. The units can be lined up with a picket boat anchored in line with the route to be taken. On the return sweep the beachmaster can use signals to keep the boats on a straight course. One method of using several boats tied together is illustrated in Figure 7.48.

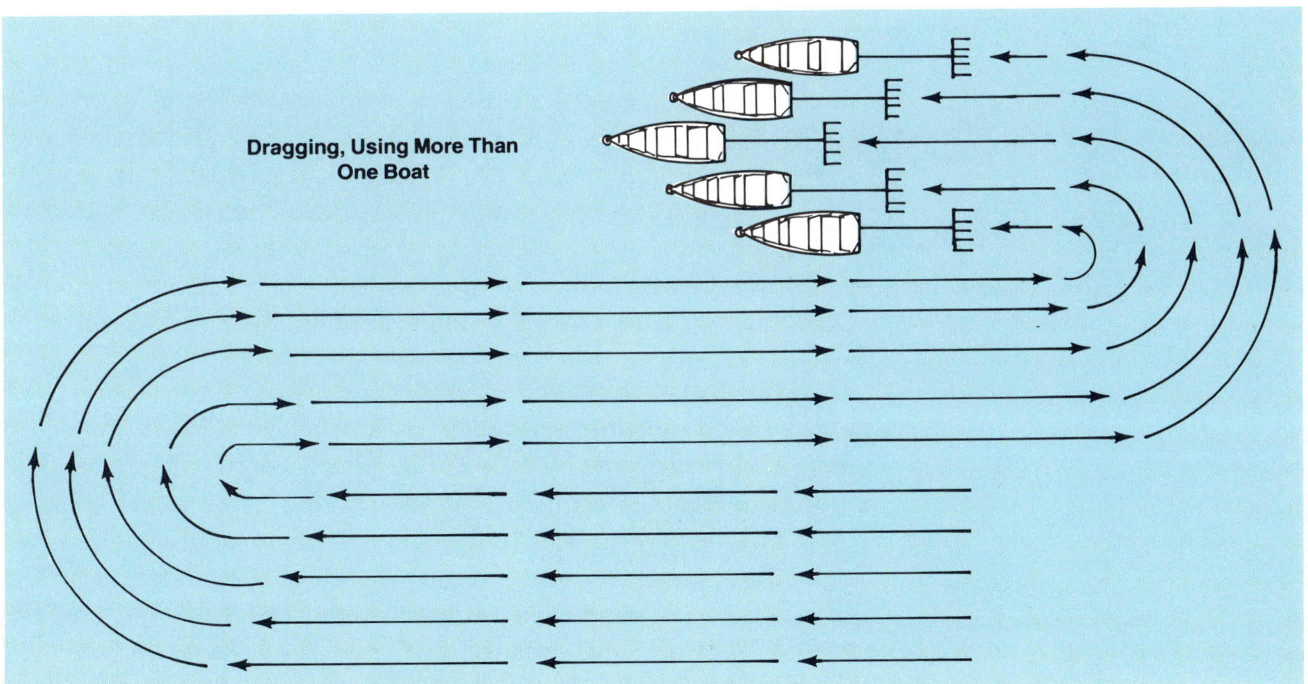

Figure 7.48 Using more than one boat for dragging.

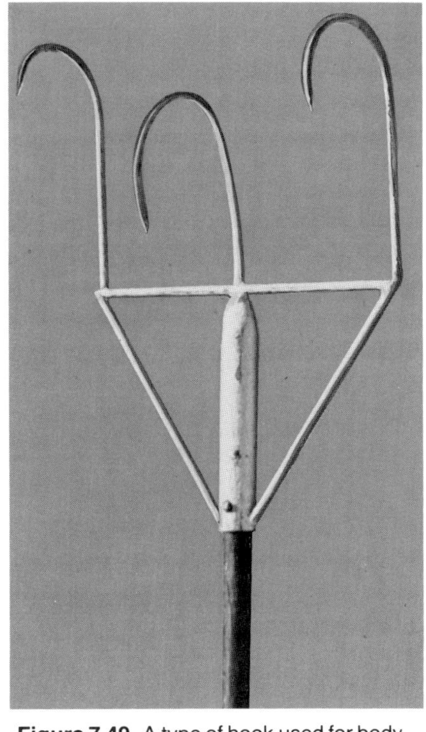

Figure 7.49 A type of hook used for body recovery.

The hooks used for drag lines or grappling devices should be sharp and not too large (Figures 7.49 - 7.51). The hooks will enter through the clothing or into the flesh of the victim. Hooks that are dull or too large often drag over a victim's body without recovering it. Lines with hooks and small weights can be thrown from the bank or from a boat. Let them reach bottom before slowly pulling them in.

When conditions prohibit using drag and grappling hooks, use a pike pole to probe if the water is shallow. Jointed light pipe with stationary hooks can also be used. Secure joints of pipe together to reach greater depths. The probing method is more successful in shallow water.

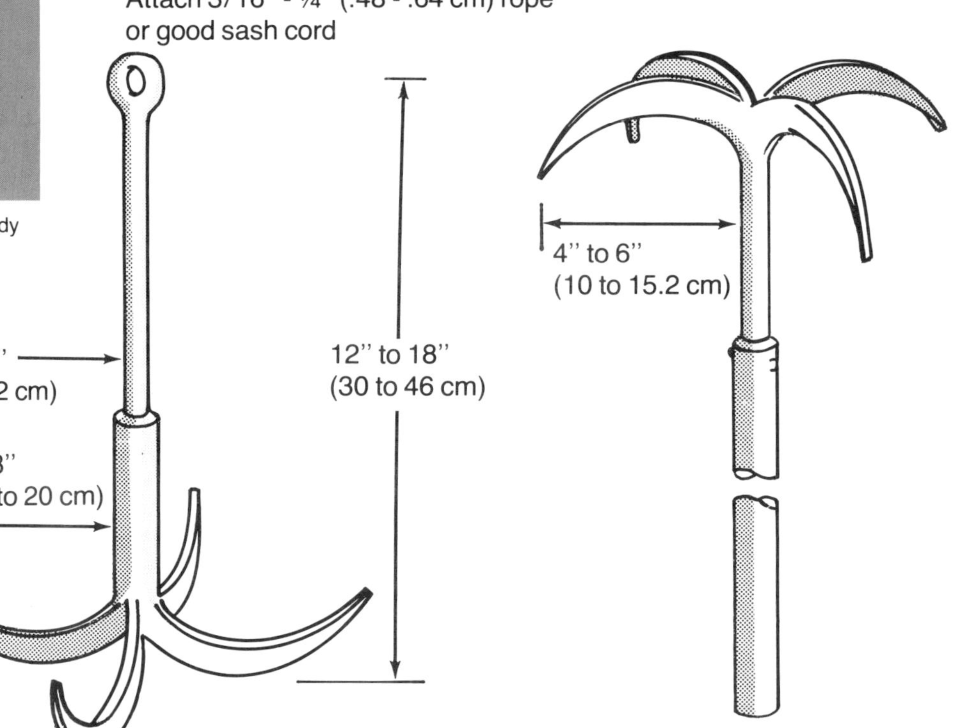

Figure 7.50 This hook is most useful for deep water.

Figure 7.51 The pike pole iron, shown here, is the only kind useful in areas of submerged stumps and similar obstacles.

Scuba divers must also work systematically. One good method is illustrated in Figures 7.52 and 7.53. Anchor a boat at the point where the victim was last seen and station another boat within a distance 1½ times the depth of the water at that point (Figure 7.54). Tie the second boat to the anchored boat with a guide line. Scuba divers should then search each side of the guide line while the second boat gradually rotates around the anchored boat. A similar principle is illustrated in Figure 7.55. Here a guide line is held by two squad members, one on each side of a body of water. The scuba divers follow the guide line on each side as the squad members on the shore advance with each crossing.

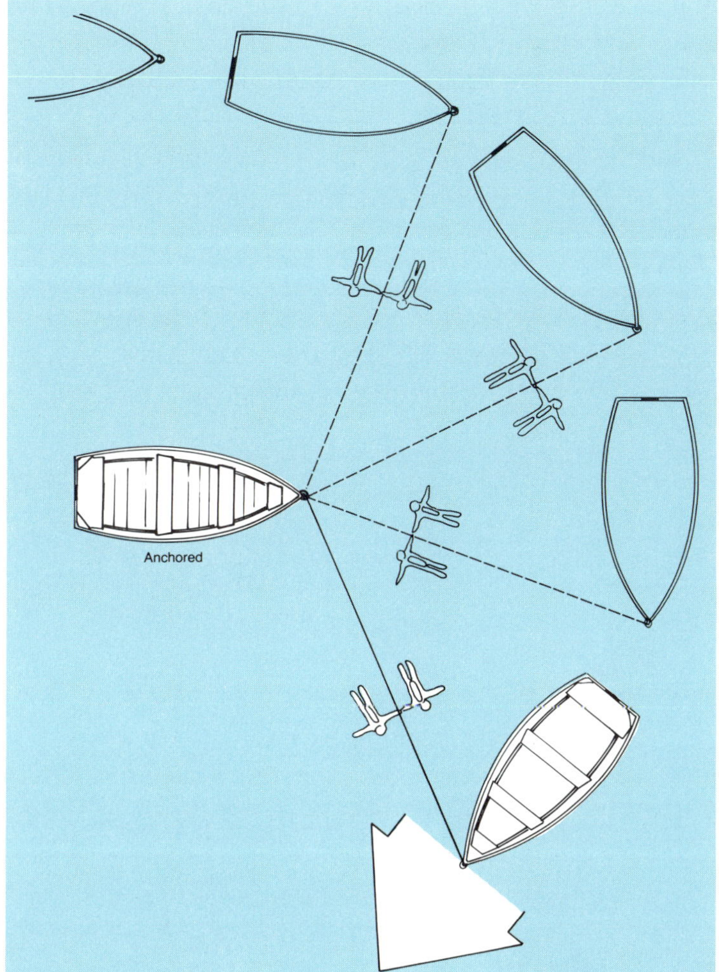

Figure 7.52 One method of making scuba sweep.

Figure 7.53 Scuba divers using guide line. *Courtesy of International Association of Dive Rescue Specialists.*

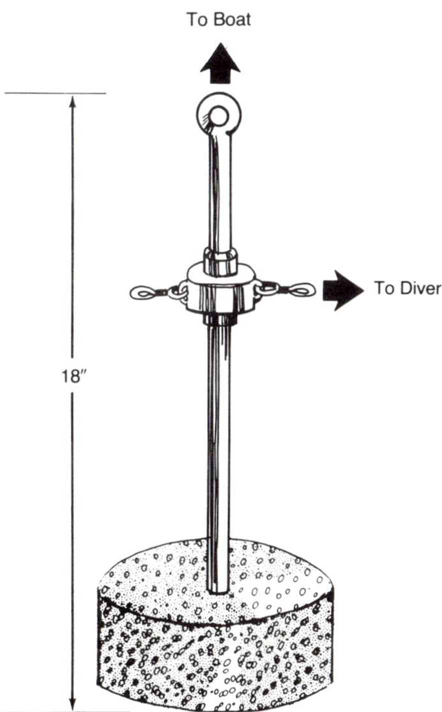

Figure 7.54 An anchoring device recommended by the International Association of Dive Rescue Specialists.

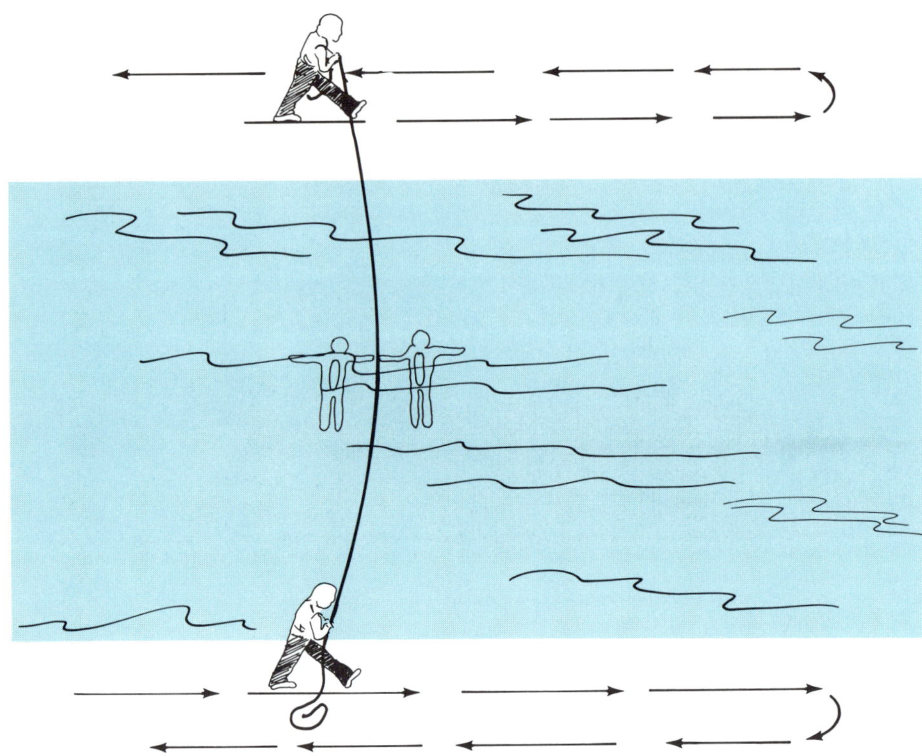

Figure 7.55 Scuba search in a running stream.

USING SCUBA EQUIPMENT

The term *skin diving* is common, but it has a broad meaning. Skin diving includes using only an eye and nose mask and possibly a snorkel and fins. But this type of skin diving has limited use and requires considerable physical endurance.

Another method of diving uses self-contained underwater breathing apparatus (scuba). Scuba equipment is essentially an open-circuit demand breathing device that consists of a tank assembly and regulator assembly. Self-contained breathing apparatus designed for fireground use should not be used in the water.

Diving without proper instruction and supervision can be dangerous. Accidents are invariably caused by a diver's mistake in judgment or technique. This section is not intended to teach rescuers how to use scuba equipment nor how to skin dive. This instruction should be given only under proper supervision by qualified instructors with approved equipment. Training in scuba diving is a special course for those who can qualify. The main objective and purpose of diver training is to teach students to think clearly under water, to make correct choices, and to act with coordination, precision, and control. Scuba diving equipment is shown in Figure 7.56.

Figure 7.56 Scuba divers getting ready for body search. *Courtesy of Montgomery Co., Md., Fire/Rescue Services.*

Cave Rescue

Chapter 8

NFPA STANDARD 1001
Fire Fighter I

4-12 Rescue

4-12.4 The fire fighter shall identify dangers of search and rescue missions in tunnels, caves, construction sites and other hazardous areas.

Reprinted by permission from NFPA Standard No. 1001, *Standard for Fire Fighter Professional Qualifications*. Copyright © 1981, National Fire Protection Association, Boston, MA.

Chapter 8
Cave Rescue

Although firefighters will often be called when someone is lost or injured in a cave, the firefighter is probably not the best person to perform the rescue. Rescue from caves must be done by persons who are familiar with the uniquely hostile environment of a cave, and who have and know how to use the special equipment that is needed. For example, typical turnout gear is of no use in caves and can be, instead, hazardous. Turnout coats and pants are too bulky for the frequently narrow passages, the firefighter's boots give unsafe footing, and the helmet is too large and cannot carry a light. The usual fire department ropes and rescue equipment are not useful in caves, because rescue from a cave is very different from rescue from a structure or from mountainous or rough terrain.

The best course for the fire department is to call the National Speleological Society's Cave Rescue Commission. To contact the commission, call Scott Air Force Base (618-256-4927) and ask for a cave rescue coordinator. Skilled rescuers can be on the way in a short time (Figure 8.1).

Figure 8.1 Call on the National Cave Rescue Commission for assistance in rescues from caves.

The following section was prepared for IFSTA by Lee Noon, national coordinator, National Cave Rescue Commission. It is included in this manual to give the firefighter some idea of the techniques and problems if called upon to assist in a cave rescue.

The fire department whose district includes caves should have a rescue plan that includes standard procedures for calling in experienced rescuers and as much information as possible about all the known caves. Information concerning the caves is best obtained from the nearest grotto (chapter) of the National Speleological Society. The members of the grotto can provide details that otherwise would take much time to learn.

TYPES OF CAVES

Solution Caves

Solution caves most common problem

Solution caves are the most commonly explored and therefore are the ones from which there are the most rescues. There are two basic kinds of passages in solution caves: phreatic and vadose. Phreatic passages are those formed when the cave was below the water table. Phreatic passages tend to be large, smooth-walled, and mostly level and dry. Vadose passages are formed by fast-moving streams above the water table. They tend to be very irregular in shape, with narrow, fluted passages that often contain active streams. These passages are more difficult to negotiate than are phreatic passages.

Lava Tube Caves

Lava tube caves are formed by lava flows on the earth's surface. The lava on the outside of the flow cools and crystalizes while the lava in the middle is still molten. This molten lava flows out of the tube, leaving a void. Occasionally, these voids are stacked atop one another as the result of successive lava flows.

The greatest problem of lava tubes is their instability. From the time the lava core flows from the tube, the tube begins to collapse. This collapse continues until the cave is filled in.

Rescuers in lava tubes need brighter lights than usual because the walls are less reflective than the walls of other types of caves.

Talus Caves

Talus caves are formed by fallen rock between ridges or at the base of mountains. Although not caves in the common sense of the word, talus caves are nonetheless caves, and persons can get lost or hurt in them. The passages of a talus cave are the voids between the fallen rocks and may have streams flowing through them. The irregularity of the passages and the instability of the surrounding rocks are the greatest problems of a talus cave.

RESCUE PROBLEMS

Six basic rescue problems

Most caves have the same six basic rescue problems: darkness, water, passage irregularities, air movement and temperature, atmosphere, and route finding. The rescuer must be prepared to overcome each of these problems if the victim and the rescuer are to come out of the cave alive.

Darkness

The darkness in a cave is absolute and rescuers must carry lights with them. The rescuer should carry three independent sources of light (Figure 8.2). The primary light should be a cap-mounted electric lamp similar to those used in mining. A cap-mounted lamp frees both hands, which is necessary for negotiating passages, handling the victim, and ascent and descent. The lamp should be able to provide light, without attention, for eight to ten hours. Spare bulbs and batteries should be carried in the rescuer's pack.

Figure 8.2 Clockwise from upper left: carbide lamp, battery packs for helmet lamp, electric lamp attached to helmet, spare electric lamp, and battery pack.

The secondary and tertiary light sources could be a good, waterproof flashlight and a candle, respectively. There should be spare bulbs and batteries for the flashlight and matches for the candle.

The candle could be used while waiting (there is a lot of waiting during cave rescues) or while repairing the other light sources.

If for some reason all light sources are lost or inoperable, the rescuer should stay in one place and not try to move around.

Carbide lamps, although often used during cave exploration, should not be used during cave rescue. The open flame is dangerous to the victim and to equipment and the fuel must be changed often.

Water

Most caves, especially those formed by water, are natural conduits for water. The temperature of the water in caves is usually about 55°F (13°C); immersion will cause major hypothermia problems within a few minutes unless the person is protected by a wetsuit.

Every year many cavers are trapped by flooding in a cave. The water can come from a sudden downpour, snow melt, or other sources, filling the passages and cutting off retreat. In most of these cases, the cavers seek a high spot downstream from a tight constriction. The constriction limits the flow of water and will often leave an air pocket.

In most cases of flooding, the rescuer must wait for the water to subside, which could take from a few hours to a few days. The rescuer must remember, once entry can be made, that the ground is saturated and even a slight rain could cause reflooding that could trap the rescuer.

Sometimes scuba diving will be necessary to rescue the trapped cavers. Such divers should be cave divers. The NCRC can get highly trained cave divers to a site quickly.

Passage Irregularities

Cave passages tend to be very irregular. They range in size from large, smooth-floored "subway tunnel" passages to very low, narrow, sinuous passages with several inches of water on the floor (Figure 8.3). Often, one must compress one's chest to negotiate these small passages. Tight passages can give claustrophobia to persons other than cavers, although these persons would probably never suffer from it in any other situation.

Figure 8.3 Many cave passages are very low and sinuous, making conventional rescue techniques inappropriate. *Courtesy of Lee Noon.*

Some passages are so small that the victim will have to stay underground until an injury heals enough that the victim can help while being taken through the cave. An alternative in such a case, if the victim's exact location is known, is the mine rescue drill operated by the Mine Safety and Health Administration's Mine Emergency Operations division. The mine rescue drill can make a 24-inch- (60.96 cm) diameter shaft through solid limestone at a rate of 50 feet (15.24 m) per day.

The rock of a cave tends to be stable. Although the floor may be covered with rocks ("breakdown"), most of the rock fell hundreds or thousands of years ago. Because of this stability, shoring and roof jacks are seldom needed.

Vertical drops in caves range from a few feet to more than 500 feet (152 m). Such drops can be negotiated safely only by persons skilled in ascending and descending fixed ropes. The rescuer must also be familiar with cable ladder and belaying (safety rope) techniques and with the use of natural and artificial anchors and their rigging. A rescue is not the time to try to learn these skills.

Cave passages can make a rescue very difficult. For example, the rescuer might have to lift the stretcher 100 feet (30 m) up a free drop and then, while hanging on a rope to the side, push the stretcher through a hole two feet (60 cm) wide and 14 inches (40 cm) high, after which the stretcher will have to pass through several hundred feet of ten-inch-high (25 cm) passage whose width never exceeds two feet (60 cm).

Air Movement and Temperature

A cave's temperature is generally the average temperature of the surrounding countryside. In the United States, most caves have temperatures between 52 and 60°F (11 and 15°C). The rocks, mud, and water will all be this temperature — eventually, so will an unprotected victim.

Hypothermia greatest danger

The low temperature inside a cave is the greatest environmental hazard to the lost or injured caver because it causes hypothermia. The caver loses body heat through conduction into the rocks, mud, and water; from general radiation; from convection as the warm barrier of air is moved from the body by air currents; and by respiration. An uninjured person can keep warm by moving about and by eating high-carbohydrate food, but an injured person is likely to have to lie on the rock, with little means of keeping warm.

Hypothermia caused by conduction, radiation, and convection can be treated simply by putting the victim into a neoprene exposure bag that covers every part of the body but the face (Figures 8.4 and 8.5). Hypothermia aggravated by respiration loss, however, is best treated with a warm-gas inhalator.

Figure 8.4 Exposure bag for hypothermia victim. *Courtesy of Lee Noon.*

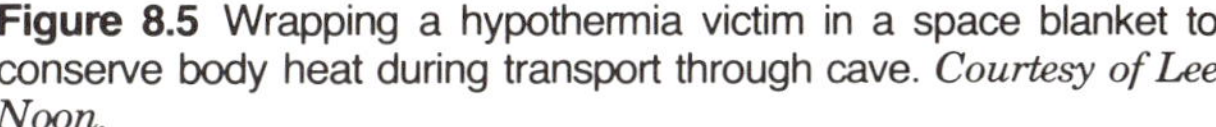

Figure 8.5 Wrapping a hypothermia victim in a space blanket to conserve body heat during transport through cave. *Courtesy of Lee Noon.*

Atmosphere

With a few notable exceptions, the air in a cave will be as pure as that of the surrounding area. Unlike mines, caves ventilate themselves very well with changes in barometric pressure. Therefore, the rescuer seldom needs ventilation blowers or breathing apparatus. Some caves, however, may contain carbon dioxide, carbon monoxide, oxides of nitrogen, or other pollutants at dangerous levels.

Caves may contain toxic gases

Carbon dioxide, when present, is usually at an acceptable level. The most common source of carbon dioxide in a cave is decaying vegetation washed in during high water. Another source of carbon dioxide is the exhalation of too many people in a small, cramped space that does not have air movement.

The carbon dioxide will pool in the lowest spot, just as water would pool. This pooling can be a problem for a trapped caver. When supplying such a caver with oxygen through a tube, the end of the tube must be below the caver's face; otherwise, the heavier carbon dioxide will "drown" the victim.

Carbon monoxide is not a natural cave pollutant but is generated by man. The most common cause is the use of small stoves to heat food and water. Another possible cause is the use of gas lanterns. A third source, of which all rescuers must be aware, is the use of gasoline engines outside the cave. This would include gasoline-engine-driven generators brought to the site for night-time outside lighting. The exhaust from these engines can be drawn into the cave if they are not both downslope and downwind from the entrance. Do not use a gasoline engine if the exhaust fumes could be carried to the cave by shifting winds.

Nitrogen dioxide and other nitrous compounds are the result of blasting in a cave. Blasting is becoming more common as cave explorers open passages too small to be passed by any other means. Rescuers also must sometimes use explosives to widen tight constrictions so a victim can be brought out. To keep the nitrous oxides as low as possible, only mining-permissible explosives should be used.

During at least two recent rescues, liquid and vapors of gasoline have been found in caves, resulting in deaths of several cave explorers and rescuers. These deaths were caused by explosion and inhalation.

If there are quantities of toxic gases in a cave, the rescuer must wear long-term, self-contained breathing apparatus such as BioMarine or Draeger oxygen rebreathers. The short-term SCBA commonly used by the fire service is more hazard than help: they allow one to get into the cave, get into trouble, and not have sufficient air to leave the cave when the alarm bell rings. (One must remember that cave rescues are protracted affairs and involve prolonged, strenuous activity.) An SCBA will be needed for the victim, also.

Route Finding

The very nature of caves makes route finding very difficult. Outside, on a mountain, for example, one can see all around and where one is going; but in a cave, one can see only as far ahead as the next bend and behind — if one can look back at all — only as far as the last bend. The route must be mapped in one's mind as progress is made.

Many caves are very complex — some with several levels of labyrinthian passages and only one or two entrances or exits (Figures 8.6 and 8.7). Because of this complexity, many rescues involve lost parties. This may be as simple as a disoriented party being just a few hours overdue or as difficult as a party being totally lost in a complex section of the cave.

Experience has shown that the best way to find parties in a cave is to send small teams on a rapid sweep through the most commonly traveled passages. The premise of this procedure is that the party is conscious and wants to be found. The first such sweep will usually find a lost party.

If the first sweep is unsuccessful, a more detailed search is made, with the party's exploration preferences taken into account: do they push low, tight passages; do they prefer climbing; did they wear wetsuits for water passages?

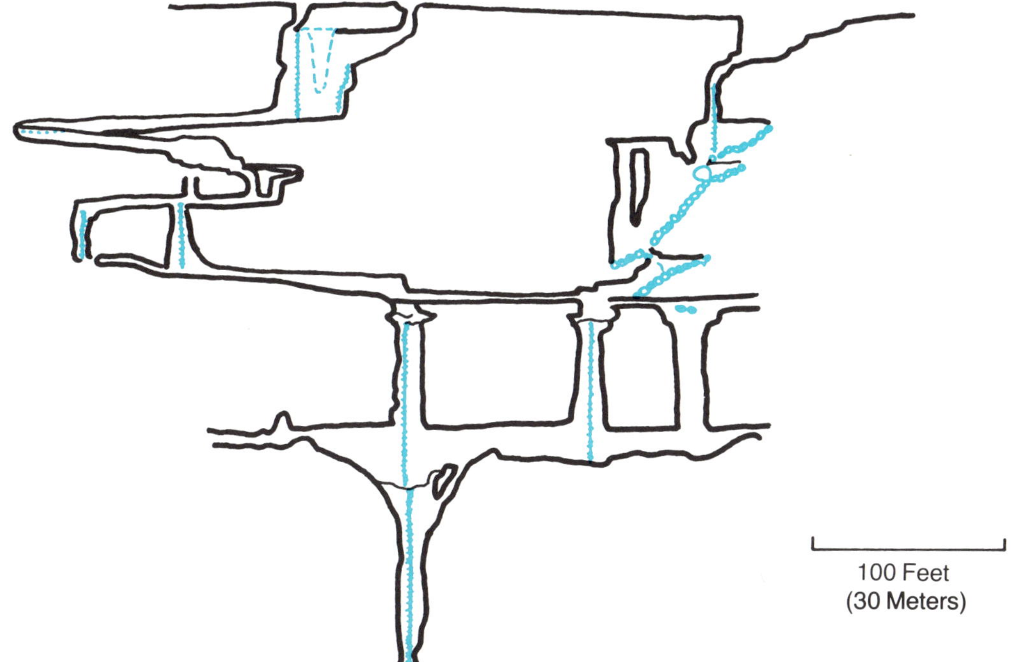

100 Feet
(30 Meters)

Figure 8.6 Cross-section of a limestone cave.

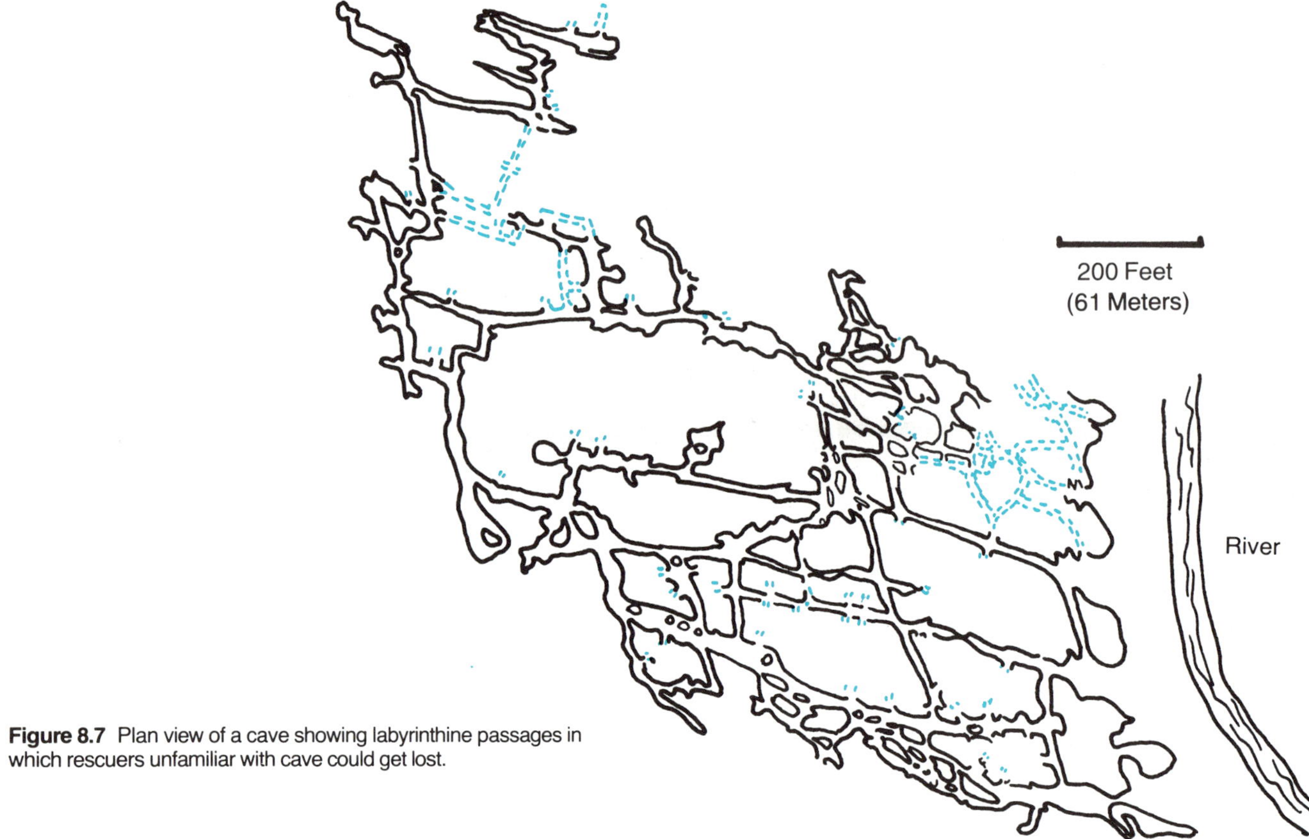

Figure 8.7 Plan view of a cave showing labyrinthine passages in which rescuers unfamiliar with cave could get lost.

The effects of joints on cave development in deep but level strata. Note: The basic passage is horizontal and the shafts are developed along joints.

One thing that must be remembered is that even a cave map of great detail might not show all the passages of the cave. Experienced cave explorers might be in passages never before discovered.

EQUIPMENT

General

Equipment for cave exploration and rescue has been developed over the years to deal with the unique environment of caves. The basic gear for horizontal travel includes coveralls, vibram-soled boots with lug tread, and a shoulder pack for miscellaneous gear such as extra lights and food. For vertical travel, the rescuer will need a rappel device such as a rappel rack for descending fixed lines, and prussik knots or mechanical ascenders for climbing fixed ropes. All of the equipment should be personally owned to be sure of its availability and condition when needed for a rescue.

The rescuer will also have to carry team equipment through the cave. Such equipment includes a stretcher, field telephone with sufficient wire (radios do not work underground), and a warm-gas inhalator. If the rescue involves vertical travel, team equipment will also include several pulleys and safety cams (Figure 8.8) and several ropes of assorted length.

Figure 8.8 Equipment for vertical work in a cave: Gibbs ascender, pulleys, and static kernmantle rope threaded through Jumar ascenders.

Stretchers

The best stretcher for cave rescue is the Niells-Robertson, with an added frame for supporting back injuries (Figures 8.9-8.11). This stretcher is not on the market in the United States, but is being handmade by a number of cave rescue teams. Another useful item is the drag sheet. Made of conveyor belt material, the drag sheet is used to drag a victim through low passages.

The Stokes basket stretcher is useful if the passages are large enough, but the passages of most caves are too small for it. Such items as the backboard, scoop, and canvas army stretcher are unsafe for use underground.

Figure 8.9 The best stretcher for cave rescue is the Niells-Robertson, shown here. *Courtesy of Lee Noon.*

Figure 8.10 Person strapped into a Niells-Robertson stretcher is immobilized to prevent further spine injury, if any, and has cranium protected from rock. *Courtesy of Lee Noon.*

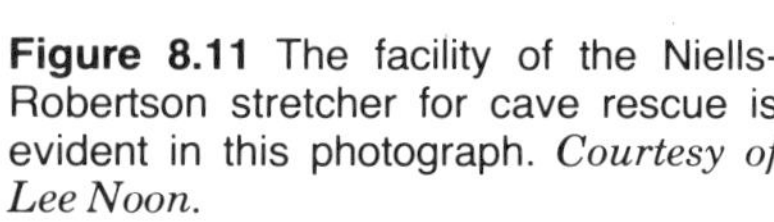

Figure 8.11 The facility of the Niells-Robertson stretcher for cave rescue is evident in this photograph. *Courtesy of Lee Noon.*

Warm-Gas Inhalator

The warm-gas inhalator, developed in the early 1970s by Evan Lloyd of Scotland, is the most efficient means of treating hypothermia in the field and should be standard equipment. Carbon dioxide is passed over soda lime, generating heat. Compressed breathing air is then passed over the hot soda lime and is warmed to about 115°F (46°C). The warmth from the breathing air is absorbed through the lungs into the bloodstream and distributed throughout the body. This method is better than warming the victim externally. External warming causes the cold-constricted superficial blood vessels to dilate, which robs the body's interior of warm blood and can be fatal.

Figure 8.12 The correct method of climbing a cable ladder. *Courtesy of Lee Noon.*

Cable Ladders

When many people must get up and down short drops (50 feet [15 m] or less), the cable ladder can be useful (Figure 8.12). The load limit of a cable ladder is very low (600 pounds [272 kg]) and can be exceeded by one climber taking a short fall. This load limit must be borne in mind. At all times, a person on a cable ladder must have a safety belay.

Ascending Equipment

Ascenders are used to climb fixed ropes and as safety ratchets when hauling a victim up a slope or sheer drop (Figure 8.13). Prussik knots* are the easiest, but are also the easiest to snarl on the rope. Jumars* are easy to use, but early models will fail if shock loaded. One of the best ascending devices is the Gibbs cam (Figure 8.14). The Gibbs cam is simple and will not fail at reasonable loads. It also has the advantage of holding well on muddy rope.

All ascenders will slide freely when no load is placed on them, but will lock when a load is applied, which makes it possible to climb ropes and to hold loads during hauling.

*See Glossary

Figure 8.13 Rappel rack and carabiners for vertical work. The locking carabiner, left, is highly recommended.

Figure 8.14 Gibbs ascenders used for climbing ropes in vertical passages.

Ropes

The only ropes that should be used for cave rescue are kernmantle, or core-and-sheath, static ropes. Laid ropes, such as are common in the fire service, should never be used. Dynamic climbing ropes should be used only when someone must do lead climbing that might result in a long fall, the need for which is seldom encountered during a rescue.

RESCUE ASSISTANCE AND TRAINING

When there is a need for rescue from a cave, contact the National Cave Rescue Commission through Scott Air Force Base. The NCRC maintains a rescue call list for teams throughout the United States. The NCRC can provide trained advisors, teams of cavers, and special diving teams.

NCRC offers training

The National Cave Rescue Commission also conducts training sessions for interested persons. For general information, write to the National Cave Rescue Commission, Cave Avenue, Huntsville, Alabama 35810.

Extrication From Vehicles

Chapter 9

NFPA STANDARD 1001
Fire Fighter II

4-12 Rescue

4-12.5 The fire fighter, operating as a member of a team, shall demonstrate the extrication of a victim from a vehicle accident.

Reprinted by permission from NFPA Standard No. 1001, *Standard for Fire Fighter Professional Qualifications.* Copyright © 1981, National Fire Protection Association, Boston, MA.

Chapter 9
Extrication From Vehicles

Extrication calls increasing

There are millions of motor vehicle accidents in the United States every year, and the number of responses by fire departments and rescue units is increasing. Some of the tools and equipment used in vehicle rescue have already been covered in this manual and will be mentioned here only in passing. But other tools and equipment necessary in extrication will be covered more thoroughly. Limited space also does not permit a review of patient and victim handling. That information can be found in IFSTA's **Fire Service First Aid Practices.**

PREPAREDNESS

Serious vehicle accidents often produce mass confusion. The rescue officer must quickly and thoroughly size up the scene to proceed with rescue operations. As with any function of the fire department, preparedness, or planning, plays a major role.

Training

The effectiveness and professionalism displayed by the fire department or rescue unit responding to vehicle accidents and how well rescuers are prepared depend on the amount and degree of training the department receives.

Hands-on training necessary

A hands-on working knowledge of all tools and equipment is vitally necessary to being properly prepared. Rescue personnel should realize that using this equipment must become second nature to them. They will be confronted with many situations where improper use of a tool, because of the closeness of a victim, could cause further injury or death. Rescuers must know how and why a tool works, when to use it, what it will do, and, most important, its limitations.

Mock accident scenes are an invaluable training aid for rescue teams. Where possible, each team member should play the roles of rescuer and victim. By taking the "victim" role, the rescuer gains a greater understanding and insight into the needs of the victim. By carrying out the duties of rescuer, firefighters increase their knowledge and experience by working around "victims" and using rescue tools and equipment.

Train frequently for local circumstances

The training program should include frequent classes on local conditions, including emphasis on special hazards such as dangerous intersections, bridges, railroad crossings and train schedules, hazardous material transport, and rush-hour times. **Firefighters should know alternate routes to reach any area.**

Because of ever changing auto designs and safety features, continuing training and flexibility of operations are musts. New federal safety laws have further compounded the problems of vehicle extrication. Rescue personnel should continually update and revise training to meet these needs. Buses, transport trucks, and other special types of vehicles present special problems to the rescuer. The training program should include classes concerning these vehicles.

Safety

Safety in operations is essential. Without the proper safety precautions no operation can be successful. Rescuers who disregard safety put victims in further danger and themselves in danger of serious injury. This applies to training or an actual emergency.

Rescuers should understand the importance of protective clothing. Vehicle accidents offer a variety of hazards. Where fire is involved, smoke, toxic gases, and extreme heat can be expected. Other hazards include glass shards, sharp metal edges, flying glass and metal, and, depending on the type of accident, dangerous chemicals or radiation. There is always the danger of tool failure or an unstable vehicle. Wear complete turnout gear including **faceshields during any operation.**

Availability of Resources

Be familiar with available aid

Rescuers should be thoroughly familiar with all resources available, including police and fire departments, ambulances, other rescue units, wreckers, CHEMTREC (see Appendix A), power companies, radiation teams, and any other agency that might be needed.

Psychological Preparedness

Being assigned to a rescue unit does not necessarily qualify anyone for rescue service. Rescuers must have a genuine feeling for other people. They must mentally prepare for the trauma of facing victims young and old who are burned beyond recognition, torn to shreds, or drowned. They must be encouraging when they feel discouraged. In short, rescuers must be psychologically prepared to do whatever is necessary for the welfare of others.

Vehicle Readiness

Keep response vehicle ready

The rescue unit must be properly maintained with a preventive maintenance program. Refuel vehicles after each run. When an emergency occurs, the unit could be running two or three hours at a time. Check the oil, battery, tires, radiator, lights, and sirens daily for proper operation. Keep records so oncoming crews know the exact operating status of the vehicle.

RESPONSE

Information

The information received when an emergency call comes in is important to the success of the operation. Information should include:

- Location of accident
- Kinds of vehicles involved
- Number of vehicles
- Condition or position of vehicles
- Number of people injured and type of injuries
- Any special hazard information
- Name of person calling and call-back number

Size Up or Assessment

Size up begins before arrival

The size up or assessment of an accident theoretically begins on the training ground, but realistically it begins when the call is received. With proper information the rescue officer can begin assessment while responding.

Location. Where streets have similar names, exact location is vital to response time. A careless dispatcher could delay rescue crews by critical minutes by not obtaining location information properly.

Kinds and number of vehicles involved. Knowing the kinds of vehicles and how many are involved gives the officer the information needed to determine the amount and type of equipment and personnel required at the scene.

Number of people injured and extent of injuries. This information lets the officer send the needed ambulances and know what to prepare for. Realize that most callers are excited, and sometimes it is hard to obtain exact information on injuries. Take this into consideration, assume that the information is correct, and act accordingly.

Special hazards. Information about special hazards lets the officer prepare for the emergency. Special hazards may include downed electrical wires (see section on electrical rescue), fire, and hazardous cargo. If fire is involved, the officer can ask that an additional unit respond with the rescue unit or ambulance. Extrication should not be attempted until the fire is controlled. If hazardous cargo is involved, firefighters must take extra measures to protect victims and rescuers. This might include evacuating the area and calling on special help such as CHEMTREC and radiation monitor teams.

All these things can be considered en route; the officer must immediately identify and correct any life-threatening hazards and determine if additional assistance is needed on arrival.

Figure 9.1 Cribbing used to stabilize an automobile. *Courtesy of Louisiana State University.*

Vehicle Stabilization

The rescuer often finds a vehicle on its side or upside down in a ravine or gully or on a hillside. Resist the temptation to rock it or push it. Often this pressure is all that is needed to tip the vehicle over, producing disastrous results to victims and rescuers alike. A vehicle in any precarious position should first be stabilized. Under no circumstances should a vehicle be tipped over with victims still inside. Stabilization may be accomplished by several means—cribbing or wedges, jacks, come-alongs, and air bags (Figures 9.1 - 9.4).

Figure 9.2 Wedges and ropes used to stabilize an automobile. *Courtesy of Elmer Reese.*

Figure 9.3 Wedges and bumper jack used to stabilize an automobile. *Courtesy of Elmer Reese.*

Figure 9.4 Air bags are quick and easy to use and will lift and stabilize vehicles of all types. *Courtesy of Jerry G. May, Beaumont, Tex., F.D.*

Cribbing can be made from two-by-fours or four-by-fours about 18 to 24 inches (45 to 61 cm) long. Cribbing should be rough, unfinished wood because painted surfaces tend to be slick when wet. Hardwood such as oak should be used if possible. Wedges can be made from the same material 12 to 18 inches (45 to 61 cm) long and 2 to 6 inches (50 to 150 mm) thick. A rope handle can be added to cribbing by drilling a hole in the block about 1½ to 2 inches (38 to 50 mm) from the end (Figure 9.5). This allows cribbing to be carried easily. For quick removal from the rescue vehicle, ends of cribbing can be painted to identify length. Cribbing is generally used to build a box crib under either or both sides of a vehicle to stabilize it, or for safety when lifting or jacking a vehicle off a victim.

Figure 9.5 Rope handles on wedges and cribbing make handling easier. *Courtesy of Elmer Reese.*

Hydraulic jacks may be used for lifting or stabilizing the vehicle, but avoid mechanical jacks. They are not ideal for this situation.

If a come-along is used for vehicle stabilization, it should be anchored to a stable point such as a tree, pole, or other vehicle. Two come-alongs on either side of the vehicle, with the hooks anchored at different places, work even better. A chain anchored to one side and a come-along to the other also works well. When anchoring the come-along to another vehicle, make sure there is no possibility of fire and that the other vehicle will not have to be moved before extrication is complete.

In an emergency several things can be used for stabilization, including the bumper jack, spare tire, the hood, and the trunk lid. When a vehicle is on its side, the spare can be placed under the wheels and the bumper jack used to jack the top up until the wheels touch the spare. Punch a hole in the roof, insert the jack hook, and work the jack. Stop when the car wheels rest firmly on the spare. The hood and trunk lid can be opened if they will rest on the ground.

Vehicles on Hills or Cliffs

Stabilize vehicles

When vehicles are on the side of a hill, attach cables and anchor them to trees, pumpers, wreckers, or any solid anchor point before attempting rescue. Sometimes moving or jarring the vehicle will send it down the hill. When the car is balanced on a cliff, removing victims could cause the vehicle to shift and tumble over. Attach anchor chains or cables where possible before attempting rescue.

Use ropes only in an emergency. Rope has a much lower breaking point than chains, cables, or nylon webbing. Rope loses its strength when a sharp bend is made. Sharp edges can cut and weaken it.

Buses and Tractor-Trailers with Air Suspension

Take extreme care when working around vehicles with air suspension systems that use rubber bellows to support and level vehicles at each wheel with air from a compressor.

When the vehicle is involved in fire or accident, the bellows may fail and the vehicle will drop suddenly to about three inches (77 mm) from the ground. Anyone working under or next to these vehicles could be hurt or killed. When jacking buses, jack or lift only at the jack point at the front or rear of the wheels. Jacking at any other place could tear the metal.

TECHNIQUES

Doors

Even after an automobile has been involved in an accident, doors may still operate normally. Before any tools are used, all doors should be tried for quick access to the victim. Concentrate on the area around the door lock where all the door and lock strength is. Once this area has been broached, the task of opening the door is easier (Figure 9.6).

Example: Start with hand tools and pry the metal apart to give power tools an opening (Figure 9.7). Use hand or power tools to spread the metal and expose the lock (Figures 9.8 and 9.9).

Figure 9.6 A hand tool especially designed for cutting during vehicle extrication. *Courtesy of Ziamatic Corp.*

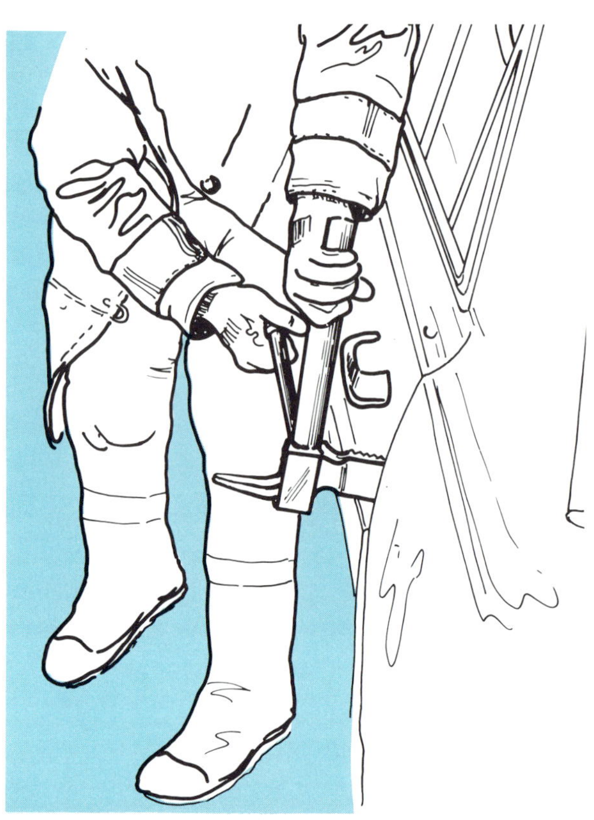

Figure 9.7 Using a Pry-Axe to pry the metal apart so power tools will have room in which to work.

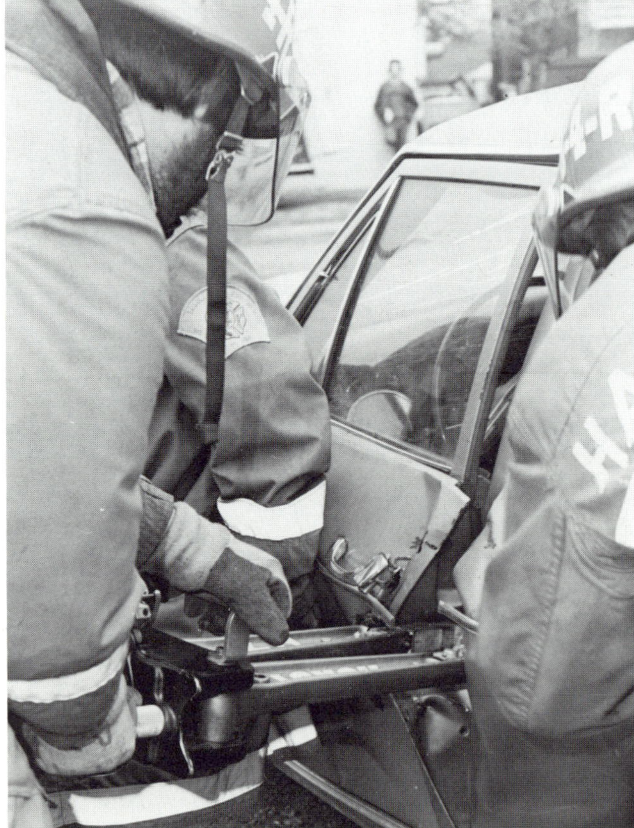

Figure 9.8 Hydraulic tool inserted near lock pries door open quickly. *Courtesy of Elmer Reese.*

Figure 9.9 High-angle view of a hydraulic spreading tool opening a door. The manufacturer recommends that two persons do this operation: one working the jaws; the other holding the door and working the pump. *Courtesy of Walker Manufacturing.*

Then concentrate efforts on breaking the lock. Place the tips of the tools as close to the lock as possible. If the tools are placed away from the lock, the force being applied has to bend more metal, reducing the amount of force actually being applied to the lock. Also try to keep the door seams from splitting, which also reduces the force being applied to the lock.

Tool placement important

Breaking the lock may not be all that is required to open the door, because metal in the hinge area or on the column may also keep the door from opening. By changing the position of the hydraulic power tools, workers can force the door wider than they can with a come-along or larger hydraulic power tools. Pull the door out of the way or remove it from the vehicle.

Windows

Take extreme care while removing glass, because glass in an open wound cannot be seen and will not show up on an X-ray. Cover all victims and rescue personnel.

There are several ways to remove side and rear windows, but the safest is to cover the window with masking tape or contact paper and use an automatic center punch, or a hammer and a screwdriver, or anything with a point. Place the pointed tool in the lower corner of the window and strike it firmly but gently. When using an automatic center punch merely release the spring (Figure 9.10). The window will shatter. Punch a small hole in the upper corner farthest from the victim and gently pull the glass out away from the victim (Figure 9.11).

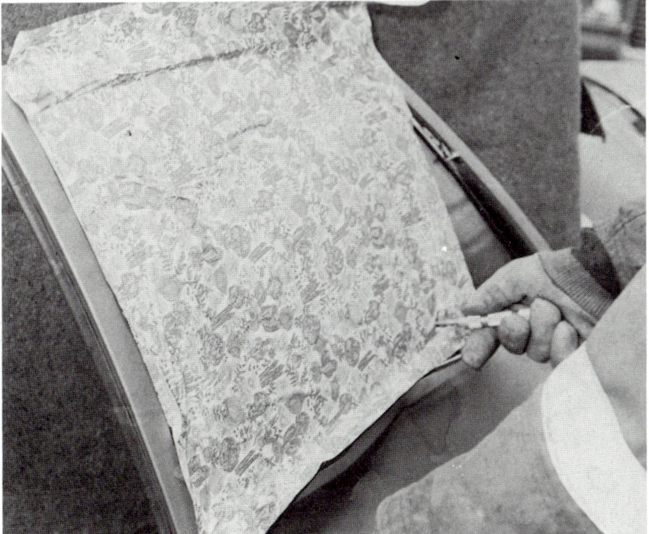

Figure 9.10 Contact paper is pressed against the window, then a pointed tool is used to shatter the glass. *Courtesy of Elmer Reese.*

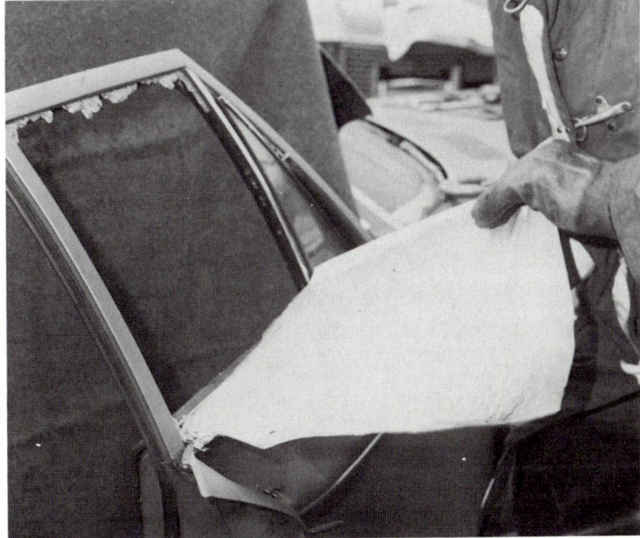

Figure 9.11 The shattered glass clings to the paper and can be removed without harming the victim. *Courtesy of Elmer Reese.*

Tops

An air chisel is a fast and efficient way to gain access through the top of the vehicle to the passenger compartment (Figure 9.12), but if a chisel is not available, a flathead axe and a hammer or some of the homemade cutting tools will also work. Take extreme caution that whatever tools are used do not penetrate too deeply into the passenger area and further injure the victim.

On upright vehicles, rescuers can cut on three or four sides of the top. Circumstances will determine what style will be most efficient. On vehicles on their sides, make a three-sided cut and leave the bottom edge intact for a smooth, flat edge to work over (Figure 9.13).

Figure 9.12 Using an air chisel is a fast way to cut through a roof. *Courtesy of Elmer Reese.*

Figure 9.13 The top is cut along three sides, bent down, and a salvage cover used for protection against jagged edges. *Courtesy of Elmer Reese.*

Floors

Vehicle floor "pans" also offer a way into the passenger area, but they are considerably harder to penetrate. Seats are attached to the floor pan, and cutting through pans with small, lightweight air chisels is a slow, difficult job. High-pressure chisels are preferred, but do not hesitate to use whatever tools are available. Remember, however, that most floors are only a single layer thick, and victims in contact with the floor can be injured during the cutting process.

Improvisation

Rescue personnel will often be at an accident scene without their rescue unit or tools. The following access procedures have been developed through necessity and imagination using materials available at the scene.

Improvised tools valuable

The tire tool and bumper jack can be used to open doors. The flat or prying side of the tire tool can be used like a chisel and the jack like a hammer to cut the metal around the lock. Then pick the lock and use the tire tool or hand to free the door.

The tire tool can also be used with the oil dipstick or radio antenna to pull the locking device when the door is locked and the passengers are unconscious. Use the tire tool to pull the glass back. Make a hook at the end of the dip stick or antenna, reach down between the glass and upholstery, and lift up on the locking mechanism.

VICTIM STABILIZATION

Stabilization a priority

Once access into the vehicle has been made, the most qualified member of the rescue team must concentrate on the welfare of the victim. Stabilization must take priority over all other rescue operations. All rescue team members must be trained in first aid procedures, and ideally they should be certified emergency medical technicians (EMTs). At the least, they should have had standard first aid and cardiopulmonary resuscitation (CPR) classes.

Primary Survey

To stabilize the victim, rescuers first conduct a primary survey to identify life-threatening injuries. Follow the steps listed below in order. (See IFSTA's **Fire Service First Aid Practices** for specific techniques.)

- Immediately clear an airway
- Restore heartbeat
- Stop severe bleeding

Spinal Stabilization

In many cases, rescuers may have to stabilize the victim's spine before proceeding to the next step.

Cervical Devices

Cervical collars, sandbags, flexible-wire splints, blankets, towels, short spine boards, and sun visors can be molded and attached to the victim's neck to stabilize the area to prevent aggravating an injury. It also makes the victim more comfortable by reducing pain. The use of cervical devices should be considered even if the victim does not complain of pain.

Short Spine Boards

A short spine board should be used when the victim is sitting upright or when a long spine board cannot be used.

Long Spine Boards

Before attempting transportation, rescuers should completely immobilize the victim, especially if they suspect spinal injuries. Always immobilize to prevent compounding an injury and causing permanent disability. Transfer the victim to a long spine board and secure. In cases where short spine boards are used, the victim must be placed on a long board for transportation after the short spine board is in place.

Orthopedic Scoop Stretcher

The scoop stretcher offers more flexibility because it can be adjusted by length and angular configuration and can be adapted to a variety of positions with less movement of the victim.

Secondary Survey and Treatment

After conducting a primary survey and stabilizing the victim's spine, rescuers should conduct a secondary survey for injuries that do not threaten life. Then administer the proper first aid.

EFFECT OF RESCUE WORK ON VICTIMS

Limit access to scene

The rescue team must take special care to reduce the psychological effect that some undesired, but necessary noise, vibration, movement, and conversation have on the victim. A certain amount of noise during a rescue operation is necessary, but it should be kept to a minimum and the victim advised and prepared to head off anxiety and fear. Vibration can add to injury and pain. Proper stabilization, immobilization, and verbal reassurance may be necessary to prevent shock or compounding injuries. Sudden movements should be avoided to prevent further injury and the psychological effect they can have on the victim. Good communication, proper training, and teamwork can go a long way toward overcoming the effects. Conversations should be limited to vital conversations only. Avoid colorful descriptions of the victim's condition, the extent of damage to the vehicle, and loud and abusive language. Conversation reassuring the victim is advisable and psychologically helpful. Limit access to the scene to trained, skilled rescue team members.

COORDINATION OF MEDICAL CARE AND EXTRICATION PROCEDURES

In vehicle rescue, total coordination of all agencies involved must be maintained. Orders must be issued through established lines of authority, because failure to follow the proper channels can result in mass confusion. The commanding officer must coordinate the entire incident from the time the emergency call is received until the incident is concluded.

All responding units, whether rescue team or support, are under the direction and supervision of the incident commander, and coordination and cooperation is a must to obtain the main objective. Support units may sometimes be involved in other than extrication procedures, all of which are important to the total picture. For example, engine companies responding should always lay and charge enough lines to control any potential hazards. In vehicle accidents the possibility of a fire is always present. Engine company crews should consider the possibilities and take the proper precautions. Law enforcement agencies responding, unless actually involved in extrication or victim care, should control traffic to prevent additional accidents and to give responding emergency vehicles ready access to and from the scene. Medical institutions should be advised about the victim's condition and be prepared to receive emergencies or assist if possible. In many cases hospitals have radio communications to advise and assist rescue personnel.

Possibility of fire always present

DISENTANGLEMENT

After gaining access, the rescuer is concerned with problems threatening the victim's life. Then the rescuer is interested in the relatively more minor injuries that might affect the victim, while at the same time seeing if the victim is trapped by the brake pedal, the steering wheel, a shifted seat, or any other mechanism. Once the victim is stabilized, the rescuer must consider disentanglement before removal.

Techniques

No two automobile accidents will be alike, but common sense and a working knowledge of rescue equipment gives the rescuer flexibility in how to do the job. Above all, do not be afraid to try something different.

DASHBOARD

There are several tools—portable power tools and cribbing, come-along and chains, and hydraulic tools—for the rescuer to use. If one fails, do not hesitate to use another tool or a combination of tools.

Start by determining what portion of the dash is trapping the victims and the best possible position for the tools. Often victims can be freed by pulling the steering wheel, but if not, push the dash with a portable power ram or a hydraulic tool. Use cribbing on the floor to keep tools from punching a hole in the floor pan.

STEERING WHEEL

To free victims trapped by steering wheels, the rescuer should first cut away a portion of the wheel to give more room to work. If this fails, use a come-along and chains, a hydraulic tool and chains, or a portable power ram to push or pull the steering column out of the way.

If using a come-along, attach a long chain to the frame member under the front of the car. Then wrap a short chain around the column and put the pulling ring through the wheel just above the center of the column. Unwind enough cable to attach the block and hook it to the pulling ring. Attach the come-along body to the pulling ring on the long chain attached to the frame. Put a cribbing block on the dash as close to the hood as possible for a better lifting angle. If the vehicle has a tilt steering wheel, be sure to wrap the chain around the column below the tilting joint. Extra cribbing may be needed on the dash for the come-along to pull the column. Do not pull the column any farther than necessary to free the victim (Figures 9.14 and 9.15).

Figure 9.14 Operator's view of using a come-along to pull a steering column. *Courtesy of Louisiana State University.*

Figure 9.15 Overhead view of using a come-along to pull a steering column. Note the use of wooden blocks. *Courtesy of Louisiana State University.*

If using a portable power ram, select the length of ram and extension necessary to get the ram under the steering column. Place cribbing on the floor so the ram base will not punch a hole in the floor. Make sure the ram does not slip off the column or the column will spring downward. Move the steering column far enough to keep it from flexing back down when the pressure is released, because the ram will have to be moved to extricate the victim with a minimum of movement.

If using a hydraulic tool such as the Hurst, put it on the hood and open its arms completely. Put the shackle attachment on the arms and lay the tool flat on the hood. Wrap the chain around the steering column, making sure the chain is below the tilting joint if there is one. Attach the other chain to a strong frame member under the front of the vehicle, bring both chains onto the hood, and attach them to the shackles on the arms of the tool (Figure 9.16). As the arms close, the column will start to bend. The rescuer may need to open the arms and repeat the procedure to move the column enough to extricate the victim. In some cases cribbing may be required to increase the leverage or build a platform for the tool to rest on without damaging the vehicle.

Figure 9.16 Using a hydraulic rescue tool to pull a steering column as the tool's jaws close. *Courtesy of Bill Maggi.*

Hydraulic tools whose jaws work best when opening may also be used to pull the steering column, as shown in Figure 9.17.

When using the tool inside the auto, be sure to protect the victim from coming in contact with the arms of the tool. Put cribbing on the floor so the tool will not tear a hole in the floor pan.

Place one arm of the tool on the cribbing block on the floor. Then open the tool slightly and put the other arm on the column. Opening the tool further pushes up the column and the dash, giving the rescuer more room to prepare and remove the victim.

When using a hydraulic cutting tool to cut either the steering wheel or the steering column, be sure to secure the unattached ends of either to prevent injury to the victim or rescuer.

Figure 9.17 Using a hydraulic rescue tool to pull a steering column as the tool's jaws open. *Courtesy of Walker Manufacturing.*

DOORS

In about 90 percent of the accidents trapping people, rescuers have only to open jammed doors to safely extricate victims. But in the other 10 percent, rescuers will have to disassemble the vehicle, trying to remove the vehicle from the victim instead of removing the victim from the vehicle. The rescuer must clear a path into the vehicle, and the most logical way is through the normal openings.

Since safety door locks were added to automobiles in 1968, opening doors after an accident has become a big problem. Rescuers should practice with the tools used by their department at every opportunity to develop the techniques they can use best and to build confidence in their tools.

Assess the extent of damage to the door and decide what tools to use. Start with prying tools and bend the metal near the lock to give power tools a place to start. A hydraulic tool can be put in the window opening to bend the door down, pulling the metal away from the locking mechanism to give a clear path for placing tools near the lock. Practice and confidence in the tools will help speed the operation.

Hand-operated hydraulic tools have two tools working together, a wedge and a spreader. Put one above the other as close to the lock as possible.

Keep both hands on tool

During spreading, rescuers should keep their hands on the tool at all times to keep the tool from jumping out and striking them. One of the rescuers operating the pump should press against the door to keep it from swinging out and hitting one of the rescuers.

Engine- or motor-powered hydraulic tools should be put in the door in the same place as a hand-operated tool. Engine- or motor-powered hydraulic tools have considerably more power, however, and make the job of opening the door easier. But try to keep the rear door seams intact. When the seams split, the metal housing the lock mechanisms flap, making it more difficult to break the lock.

Once the lock has been broken, power tools may still be needed to open the door. Place the tools so they will force the door outward. Then the rescuer can decide whether the door needs to be bent backward out of the way or removed.

Use a come-along to bend the door back out of the way. Wrap one chain around the door near the handle (Figure 9.18). Take the other chain under the car and bring the chain up over the fender. Unwind the cable and attach it to the pulling ring on the chain around the door. Attach the other chain to the body of the come-along and put cribbing under it to stabilize it. Then the door can be pulled back against the front fender, getting it completely out of the way (Figure 9.19). The same techniques work on rear doors.

Hydraulic tools are also used to move doors out of the way. They are capable of completely removing the door from the vehicle. Put the tool near the hinge with one arm against the door and the other arm against the body. Spreading the arms of the tool will break the hinge and get the door completely out of the way. Unlike hand-operated hydraulic tools, power-driven hydraulic tools can be operated by one person, freeing other personnel for other functions.

DOORPOSTS

Doorposts can be pulled out of the way with a come-along or a hydraulic tool. Before using a come-along, cut the bottom and, if necessary, the top of the post with an air chisel or hydraulic tool

Figure 9.18 Method of wrapping a chain around the door for pulling it out of the way. *Courtesy of Elmer Reese.*

Figure 9.19 Connection of come-along to frame and door for pulling door out of the way. *Courtesy of Elmer Reese.*

(Figure 9.20), and pull the post out of the way. Hook one chain to the front or rear frame and wrap the other chain around the post. Then connect the chains to the come-along and pull the post out of the way.

If using a hydraulic tool, put one arm against the seat track as close to the floor as possible. Angle the other arm against the post and push it out. The tool can also be put against the floor pan, but first remove the carpet so the tool will push directly against metal, reducing chances the tool will slip.

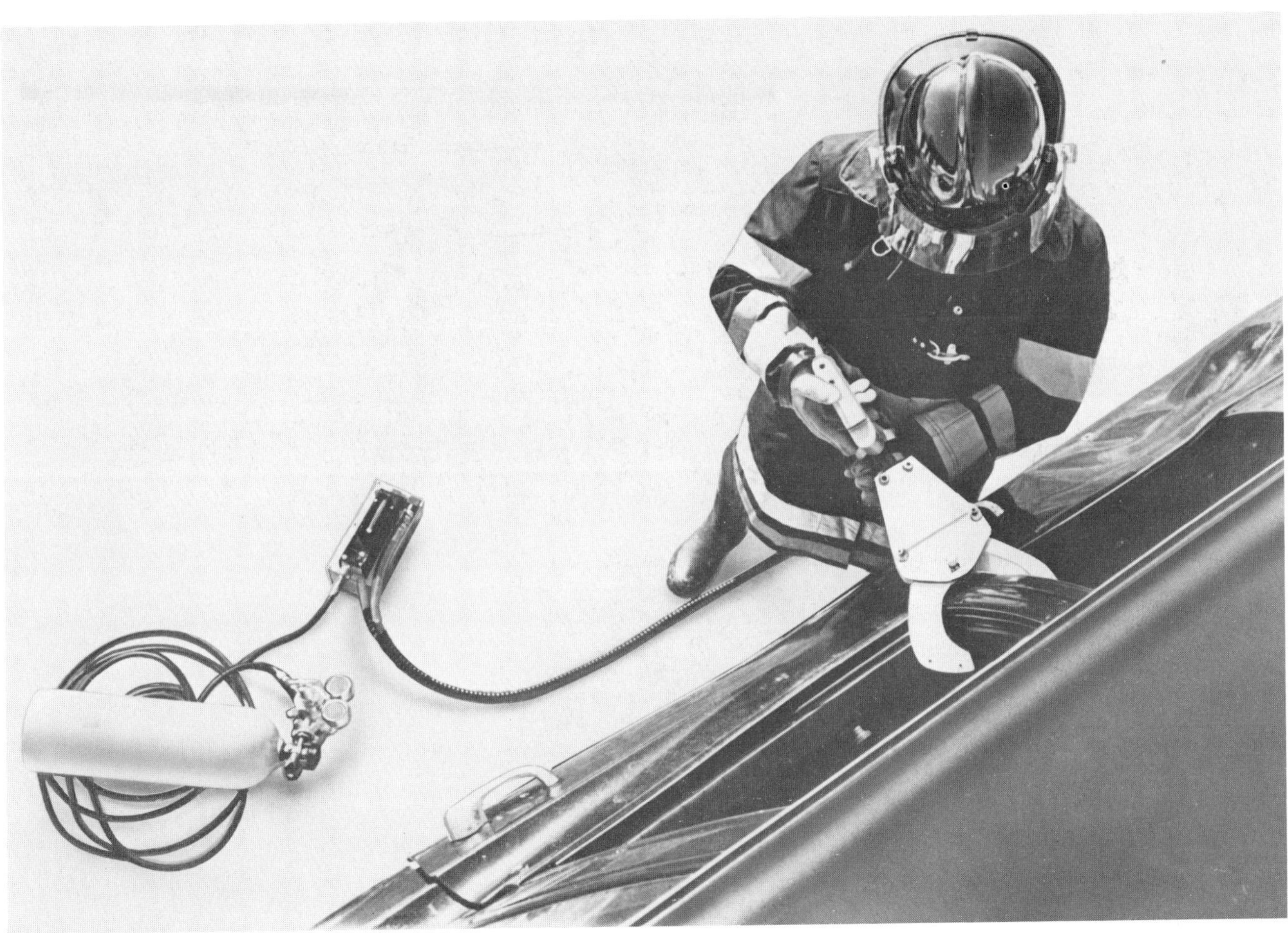

Figure 9.20 Sometimes the doorpost must be cut and pulled away so the victim can be reached or removed. *Courtesy of H. K. Porter.*

WINDSHIELDS

Windshields offer the rescuer a problem without an easy solution. Because of their construction, windshields are difficult to cut, and when people have had their heads forced through a windshield, rescuers are faced with a delicate extrication problem. The procedures listed below should be modified if necessary.

When a person's head is forced through a windshield, rescuers must do more than pull the victim back through. The windshield's plastic inner layer will let the head pass through, but not without severe injury. Sharp edges of glass may be pressing against carotid arteries.

Before doing anything, stabilize the car with cribbing or other available material to keep the vehicle from moving and aggravating the victim's injuries. Then rescuers must stabilize the victim so the head is immobilized. Afterward, it's a matter of clearing away enough glass so the victim can be removed. Cover the victim's eyes to protect them from glass and work a towel, blanket, or sheet between the neck and the windshield (Figures 9.21 and 9.22). Next, start pulling the glass from around the neck and cutting the plastic inner layer with a knife. (Figure 9.23). This is a slow procedure and there is no quick way to do it safely. When enough glass has been removed, apply a cervical collar or a backboard. Then, working as a team, move the victim away from the windshield and treat the injuries.

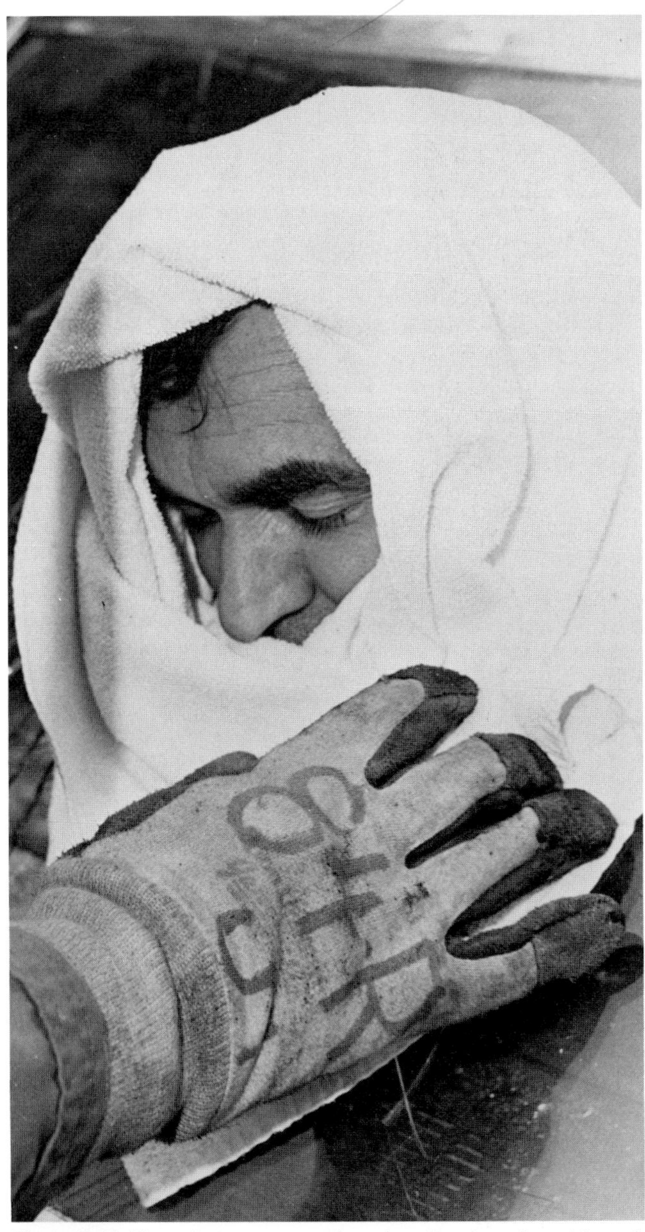

Figure 9.21 To extricate a person whose head has been forced through the windshield, first protect the person from further injury by working a towel or similar material between the neck and glass. *Courtesy of Elmer Reese.*

Figure 9.22 View from interior: protecting the victim's neck from further injury. *Courtesy of Elmer Reese.*

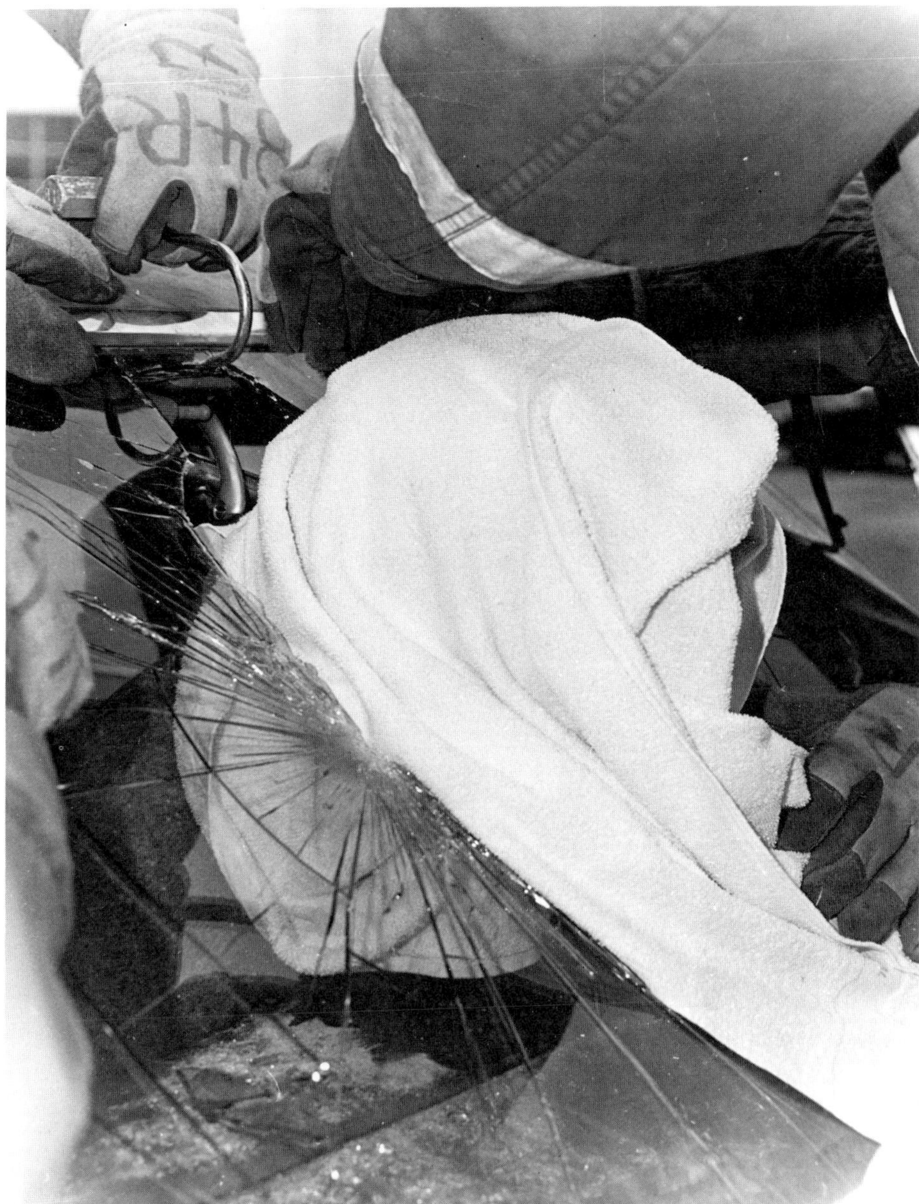

Figure 9.23 When the victim has been protected from further injury from windshield glass, widen the hole so the head can be extricated safely. *Courtesy of Elmer Reese.*

PEDALS

Sometimes a driver's foot might be trapped under the brake or clutch pedal. Use a hand-operated or engine-powered hydraulic tool to lift the pedal from the foot. If there is not enough room to use a tool, wrap a chain or rope around the pedal and carry the chain or rope to either side, depending on how the foot is trapped. Then wrap the end around the door in the direction that the pedal is to be pulled. Two rescuers can pull on the door to move the pedal. Make sure that a rescuer stabilizes the victim's leg to prevent further injury. If the door cannot be used or if the pedal needs to be moved upward, attach a chain or rope to the pedal and pull it up through the removed windshield.

SEATS

In many accidents, the victim will be caught between the seat and the dash or the firewall. To complete the victim survey, the rescuer must displace the seat.

Try before you pry

First, try to move the seat with the manual seat adjustment. If the seat and track are not damaged, it might be possible to move the seat back easily. Rescuers on each side of the seat can push it back as one of them works the lever. Power seats may still operate if the battery cable is still connected. If these methods fail, then there are several other methods that can be tried.

To use a portable power tool, attach the extensions to the hydraulic ram and place the unit between the door frame or floorboard and the bottom corner of the seat. Use cribbing at the door frame or floorboard to prevent slipping or puncturing the floorboard. Pump the unit to push the seat back on the track and use the spreader to pry the seat up and off the track.

The hydraulic rescue tool can be used in three ways. One jaw can go against the doorpost and one against the seat to push the seat back. Or, the arms can go under the seat with one jaw on the floor and one under the seat to lift it away from the track. A third method is to open the jaws and attach a cable or chain from one jaw to the seat and attach another cable or chain from the other jaw to something solid. Then close the jaws.

The come-along works by hooking the hook on the cable to a chain attached to the seat. Attach another chain to the rear bumper or frame. Adjust the chain so the winch is on the trunk or side of the quarter panel. Then operate the handle and pull the seat back.

IMPALED VICTIMS

Many injuries in motor vehicle accidents occur when objects impale the body. Only physicians should remove impaling objects, so the victim may have to be moved with the object still impaled.

Hand tools best

Often the object must be cut or pried loose from the vehicle, an operation that requires preparation and consideration for the victim. Use hand tools, whenever possible, rather than power tools. There is less vibration, lashing, or movement of the object.

Dress and bandage the wound around the impaled object to secure and immobilize it. Sometimes sandbags can be used to prop up or immobilize the object.

PREPARATION FOR REMOVAL

Before a rescuer can begin removing or transferring a victim, the team member responsible for victim care must be sure the patient is ready. If the victim is conscious, coherent, and only slightly injured, then the victim may not need to be packaged for removal to the ambulance.

"Packaging" is a term that means that wounds have been dressed and bandaged, fractures have been splinted, and the victim's body has been immobilized to reduce the possibility of further injury. Proper packaging protects the victim and aids and facilitates the victim's removal. See IFSTA's **Fire Service First Aid Practices** for the materials and techniques needed.

OPENINGS FOR EGRESS

When the victim has been properly packaged and is ready for removal, be sure to cover sharp edges that can cut rescuers and victims. Openings should be widened and edges padded with blankets or fire hose that has been split and prepared beforehand. Openings should be wide enough so the victim can be removed as smoothly as possible with no jerking or sudden movements.

Openings should be wide, padded

SUMMARY

Choose the easiest route available to gain access to a vehicle. Try to open the doors normally, but if they are jammed, the window would be the next logical choice. If windows are broken in the accident and the frame is not bent, the rescuer has an immediate route to the victim. If not, try the rear window. The rescuer here has a large opening and glass does not fall on the victim as readily as from a side window. Remember that the primary objective is to gain access and stabilize and protect the victim from further injury from sparks, glass, metal, and extrication tools.

The Firefighter in Charge

Chapter 10

Chapter 10

The Firefighter In Charge

PREPARATION FOR TAKING CHARGE

Anyone is potential incident commander

Although every firefighter may not be a potential officer, there is the chance that a firefighter or junior officer will have to take charge of an incident, regardless of rank. Each firefighter must be prepared to accept this responsibility when it occurs.

The major functions of the incident commander are making decisions and allocating resources. This calls for a knowledge foundation based on study and training. Study is essentially classroom work and outside reading, while training involves learning techniques and performing evolutions and simulations. This knowledge will be tempered with experiences from the past and the application of common sense during decision making. The firefighter who has worked several rescue calls under a competent officer will have amassed an experience base to call upon when placed in a command role. This experience of past actions should enable the use of sound judgment in operations that leads to issuing sensible orders for the successful conclusion of the operation.

A firefighter thoroughly trained and experienced in rescue techniques will know what can be done with the resources on the scene or available. This knowledge of the capabilities of both equipment and personnel will improve credibility and strengthen the leadership role.

A potential incident commander should make an honest self-appraisal. The knowledge of one's strengths and weaknesses will assist in decision making and lead to improved outcomes. This self-knowledge will allow the delegation of those tasks that the commander may be weak in performing. Strive to improve weak areas while continuing to hone strong ones.

Know the Requirements of the Job

When in charge of a rescue, there is a different set of responsibilities. Circumstances might require "getting in the thick of it," but, generally speaking, remember that the primary purpose is to manage and command, not act. Choose an effective command position and stay there (Figure 10.1). Identify the location for those needing to know.

The responsibility for an efficient rescue rests with the decision maker. Accept the responsibility while delegating details. Limit orders to single direct commands of what is to be done. Leave the how it will be accomplished to your fellow firefighters.

Figure 10.1 Effective incident leaders will establish a command position, identify the location, and direct the operations from there.

Learn to Behave Like a Leader

The person in charge of a rescue is a leader. Effective leaders give positive support to their subordinates when things go wrong. When the leader supports the group, the group supports the leader. The emergency scene is not the place to teach or discipline. Off-the-scene discipline is done so the wrongdoer still has pride intact. Assignments on the scene are made on the basis of necessity, not as punishment.

Effective leaders know they can make mistakes but they are able to realize when they have made one and will change tactics or strategy to fit the situation, not to fit wishful thinking. They will not blame someone else or bull ahead with the mistake. Also, the mistakes of an effective leader are new ones, not repetitions of previous ones.

The effective leader is calm in the face of all the crises at the scene (Figure 10.2). This calmness is nearly dispassionate, but there is still sensitivity to the needs and situations of subordinates and victims.

The effective leader is self-assured. This self-assurance comes from knowledge of self, others, and the right thing to do. When one has this knowledge base one can make the right decision at the right time. This is important because the person in charge is more than anything else a decision maker.

Figure 10.2 Remain calm during the situation, indicating control, but do not lose sensitivity.

DECISION MAKING

Making the decisions for a rescue operation is the same as making the decisions in other areas of one's life. The problem is defined, an objective is set, facts are gathered, the facts are analyzed using knowledge and experience, alternative solutions are considered, a solution is chosen and acted upon, and the results of that solution are evaluated. If the chosen solution turns out to be unsatisfactory, another alternative is chosen to replace it.

The human mind, however, does not work in such a linear way, which is why many decisions can be made quickly. The reason the process is outlined here as a linear process is to remind the potential person in charge that decisions at an emergency scene should not be made too quickly, that the impulse to act immediately at an emergency scene must be tempered with reason. When rational thought is abandoned, irrational decisions are likely to be made that will let the situation get out of control.

There are several decision-making memory devices for reminding the person in charge how to use linear reason at an emergency scene. One of these devices uses the acronym IFSTA:

I*dentify the problem*

F*orm an objective based on known information and resources*

S*elect one or more alternatives from the available options*

T*ake appropriate action*

A*nalyze*

I*dentify the problem.* Identifying the problem is the result of size up. Size up begins with the dispatch and approach: location of incident (highway, factory, home), reported nature of incident (vehicle accident, industrial accident, fall from ladder), and reported nature of injury (bleeding, limb caught in machine, fracture).

Size up continues upon arrival at the scene (Figure 10.3). Determine the facts — remember to distinguish between facts and conjecture. This is also the time to keep the crew in check —

Figure 10.3 As size up continues, the incident commander must identify the problem and determine all the facts affecting the situation. *Courtesy of Montgomery County Md., Fire/Rescue Services.*

the first test of training, preparation, and leadership. Emergency response crews, including the person in charge, are action oriented and must remember or be reminded to approach a situation cautiously until the person in charge has all the necessary information for making a decision. Immediate action can compound more often than alleviate a problem.

Remember to identify the major problem, not just some of the accompanying circumstances.

Form an objective based on known information and resources. In general, the goal will be to bring the incident to an expeditious and safe conclusion with the least possible risk to persons and property. A suitable objective(s) to attain this goal must be set. Be sure that elementary (needed) information has been obtained to avoid premature commitment. This will help the person in charge make "open-ended" decisions that can be expanded, built upon, or changed, rather than "dead end" decisions that cannot be changed expeditiously. With the requisite information, the person in charge can form an objective based on knowledge, experience, and consideration of available personnel and equipment.

This is the analysis step of decision making. Factors such as time, equipment, and personnel are analyzed to decide upon a realistic objective. The objective might be long term or short term, depending on the magnitude and nature of the incident.

Select one or more alternatives from the available options. Analysis of the situation and forming objectives will give the person in charge a number of alternative actions to meet the goal. This is a crucial step in making decisions because here is where the plan of action is determined. Remember when planning a course of action that the fastest way is not always the best way.

The person in charge must also rely on the ability to make decisions based on information transferred by other personnel. One should not have to personally verify the accuracy of the information passed on (Figure 10.4).

Figure 10.4 Use the information supplied by line officers as a basis for making decisions.

Once a decision has been made, keep the other alternatives in mind. Changes are a fact of life. Equipment can fail, the victim's condition can worsen, the weather can turn foul. Try to anticipate changes and anything that could go wrong and have alternative plans ready. Never let the situation force a decision. Forecast conditions, evaluate the potential of the situation, and formulate decisions and alternatives to stay ahead of any changes.

When making a decision remember that there are times when the crew can vote and times when it cannot.

T*ake appropriate action.* This is where the chosen alternative is put into action. Stay available — communication and coordination are necessary, especially with large-scale rescues.

Orders must be clear and concise so they are understood by all parties and all parties know their responsibilities.

A*nalyze.* The progress of the implemented plan of action must be closely and continuously monitored by the person in charge. Is progress as planned? Is progress satisfactory? (The rescue might not be progressing as planned but still be satisfactory.)

The implemented plan of action could have three negative results: new variables introduced, situation worsens, no effect. When any of these happen, the person in charge must choose another alternative to meet the goal. The person in charge cannot hesitate to make new plans. Personal pride has no place at an emergency scene.

Problems That Affect Decision Making

Rescues can put a person under strong psychological stress. This is especially true when the rescue situation is uncommon or particularly horrifying. Some of the more common problems a person in command could have and must learn to handle are functional fixity, approach-avoidance, peer pressure, and urgency.

FUNCTIONAL FIXITY

The imagination and creativity of a person under pressure can disappear. When this happens, the person becomes a victim of functional fixity. Functional fixity is the tendency to use an object only for its usually defined purpose. For example, a pike pole is usually used for pulling a ceiling and a salvage cover is used to protect building contents. A person with functional fixity will use them only for these purposes. Another firefighter, however, could see the possibility of using the pike pole and salvage cover or blanket for an improvised stretcher (Figure 10.5).

Functional fixity can be extended to operations. An example is the rescue operation where a rescue attempt from an auto is made by prying or cutting apart the wreckage to extricate the victim when the window glass could easily be removed to accomplish the same objective (Figure 10.6).

Figure 10.5 Be innovative during rescues. Ingenuity will preclude functional fixity.

Figure 10.6 This window may have been removed to free the victim rather than forcing the door. *Courtesy of Elmer Reese.*

APPROACH-AVOIDANCE

Approach-avoidance, an inner conflict in the person in command, causes an inability to make a decision. The inner conflict is caused whenever the situation is such that any solution to the problem is unsatisfactory. For example, a victim whose hand is caught in a machine must be freed (approach), but extrication procedures could alleviate pressure on an artery and the victim might bleed to death before being completely freed (avoidance).

The person in charge is caught in a dilemma (Figure 10.7). The dilemma causes anxiety, which will tend to make the person try to escape from the conflict causing the anxiety.

A good commander knows there is no escape from the conflict and will make a careful evaluation, then decide on the best course of action and accept the consequences.

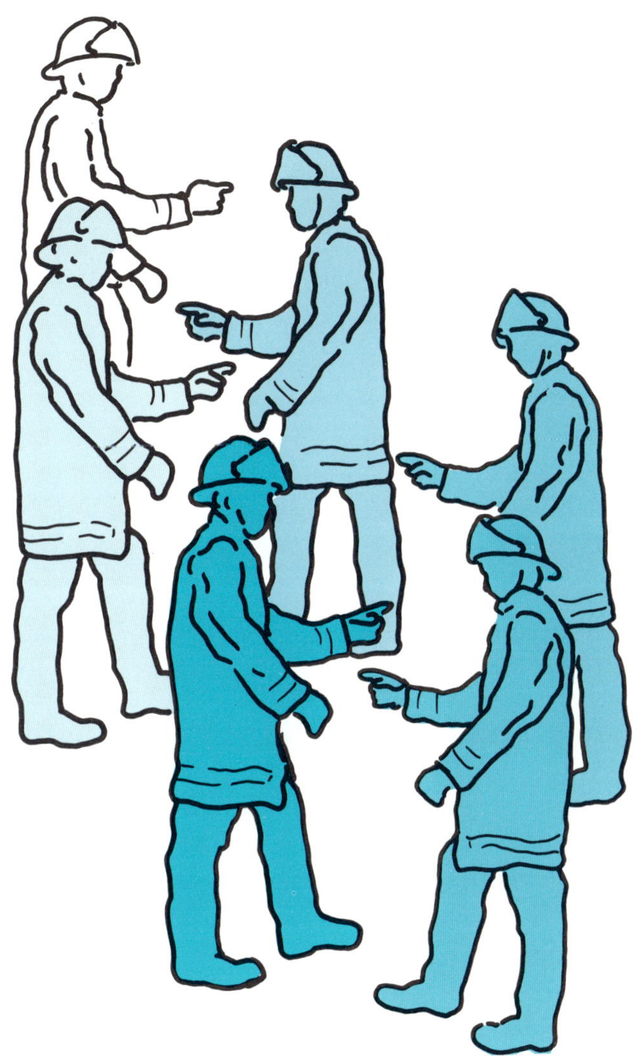

Figure 10.7 Face approach-avoidance conflicts using all available resources to develop the best course of action.

PEER PRESSURE

William Graham once wrote: "Are leaders more independent than other people, or are they adept at following their followers?" In any situation a person in command is usually directing persons among whom there is at least one who is just as experienced and just as capable of being in command. Even if there are no such persons in the group, there will be some who think they are such persons. In any case, the leader is vulnerable to peer pressure, and the leader's decisions can be affected by it.

Firefighters are just as socialized as members of any other group, if not more so. Socialized persons want to conform. Because of this, a person in charge at a rescue will sometimes make

Figure 10.8 Firefighters in charge of rescue situations must overcome the decision-making problems.

decisions based not on personal experience and good judgment but on what is thought will be accepted by the group. A person in charge must try to fight such peer pressure. If the group challenges a decision, reevaluate the decision, and, when correct, stick to it.

URGENCY

The necessity of quick decisions at a rescue scene may tempt the inexperienced leader to make decisions "from the hip." It must be understood that there will be little time to evaluate all the alternatives before making a decision. Decisions should, however, still be made by using a systematic approach utilizing the knowledge acquired through training and exerting the confidence it brings. Do not succumb to the pressures of the urgent situation and make a wrong snap decision. Also, do not fall into the trap of procrastination, thinking that no decision is a good decision.

Effects of Decision-Making Problems

The problems of functional fixity, approach-avoidance, peer pressure, and urgency will, if handled poorly, result in effects that undermine one's authority and control, incapacitate one's logic, and only make the situation worse for all involved, including the victim (Figure 10.8)

DOMINEERING ATTITUDE

A person in command is most likely to become domineering when unsure of the decisions made or when unsure of the ability to handle the situation. The domineering leader tries to control every little facet of the operation. This person will insist on strict obedience to orders and becomes easily upset with persons who do not do exactly as the person in command thinks they should (Figure 10.9). This results in a person in charge who flies off the handle and shouts such things as "I told you to use *two,* not *three!*" and "Use your common sense . . . Not *that* way!" The domineering person in charge is irrational and only adds to the confusion, doing much more harm than good.

Figure 10.9 Do not become domineering; subordinates are intelligent human beings.

AVOIDANCE

The person in charge might try to avoid being in charge, especially when faced with an approach-avoidance dilemma. This person will deal with a problem that is easier to solve and be ostensibly in charge of the whole situation while hoping that someone else will take responsibility for the dilemma. This technique works: someone else *will* take over, but the person who was in charge no longer is in charge — and might never be again.

INAPPROPRIATE RESPONSE

Anxiety can lead to inappropriate responses to the situation. The person in charge and subordinates might try to hide their fear or revulsion by joking, laughing, getting angry, or rationalizing the problem away. Such statements as "Anyone dumb enough to get his hand caught in a machine deserves what he gets" might relieve the rescuers' tension and anxiety, but it seldom relieves the victim's.

THE PARTS OF A RESCUE OPERATION

After size up and decision making, every rescue has in common the eight parts listed below. Each of these is overseen by the person in command. Remember that the first four are often done simultaneously.

1. Extricate victim from physical restraint. Physical restraint means that the victim cannot move autonomously from the area of danger. Examples are the victim caught in a machine in an industrial accident, a victim injured or uninjured in a pit or on the side of a cliff, and a victim pinned in an automobile (Figure 10.10).

Figure 10.10 Disentanglement is a primary rescue operation. *Courtesy of Vetter Systems.*

2. Get the victim out of immediate danger. Extricating the victim from physical restraint could also get the victim out of immediate danger, but there are cases where extrication alone will not do so. Examples are a victim pinned in an automobile when flammable liquids are

widely pooled as shown in Figure 10.11 (the victim must be moved from the immediate area as well as extricated), and a victim in a building when a floor, a wall, or debris in the building cannot be immediately stabilized.

Figure 10.11 Remove victims to safe areas directly.

3. Determine the nature and extent of injuries. Ideally, the victim's injuries are examined and dealt with as well as possible before either extrication or moving, but there will be times when circumstances dictate otherwise. When the incident has multiple casualties, triage must be performed.
4. Stabilize the victim and administer medical assistance (Figure 10.12). Again, ideally, much of this is done before extrication or movement. Life support is continued after extrication and movement from the immediate area.

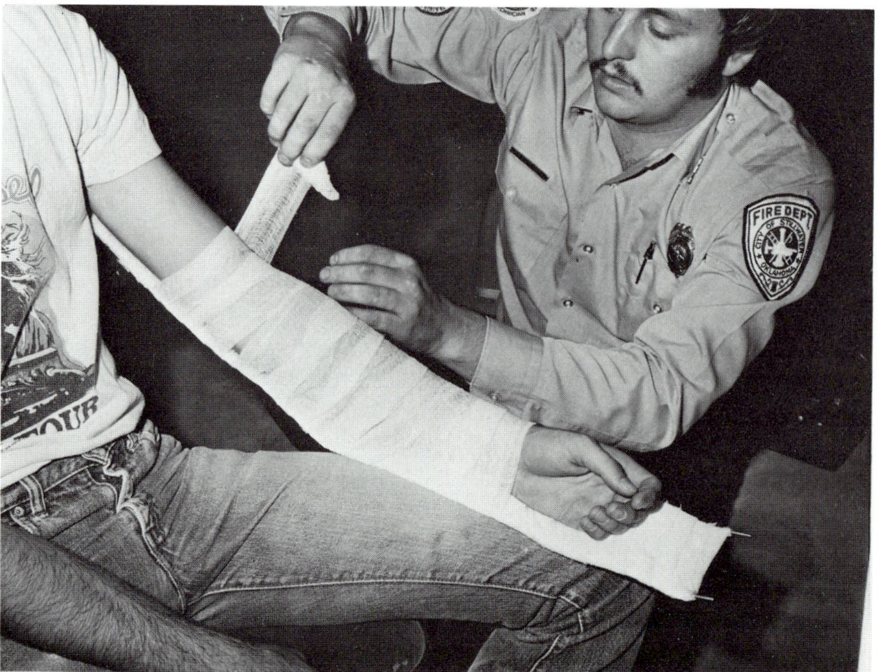

Figure 10.12 Triage, stabilize, and administer medical assistance.

5. Transport victim to medical facility (Figure 10.13). Transportation should have been called for as soon as the extent and nature of injuries were determined, sometimes before.

Figure 10.13 Transport the victim to a medical facility for treatment. *Courtesy of Memphis, Tenn., Fire Dept.*

6. Stabilize emergency scene. Make sure the scene of the rescue is made safe. This entails such things as covering gasoline spills (Figure 10.14), shoring buildings, covering holes, erecting barricades, and stationing fire or police personnel as guards.

Figure 10.14 Perform the necessary actions to stabilize the safety of the scene.

7. Wrap up. Returning emergency personnel, vehicles, and equipment to service is part of wrap up. Inventory equipment, replace what has been expended, and repair what has been broken. On return to quarters, perform standard preventive maintenance on vehicles. Write the necessary reports.

Postincident critique important

8. Postincident critique. This is the part of emergency operations that is too frequently overlooked. Constructive criticism is not easy to give or take. All too often what one person calls constructive criticism is seen as blame laying by another. After two or three well-conducted sessions, however, all involved will see the worth of the postincident critique and personalities are unlikely to enter the discussion.

As the person in charge, do not be afraid to admit your own mistakes (after all, you are not going to make them again). Your subordinates will learn from them as well as you do. They will also learn that you are an honest human being. (A word of caution: Remember the difference between admitting a mistake and being self-deprecating.)

Triage

When there are multiple casualties, triage must be performed. Depending on the nature of the incident, triage may be done where the victims are, or the victims may have to be moved beforehand. Triage is a tool for the rescuer to decide which victims will receive immediate treatment and which will have to wait awhile. The ranking of victims is:

1. Highest priority
   - Airway and breathing difficulties
   - Cardiac arrest
   - Uncontrolled or suspected severe bleeding
   - Severe head injuries
   - Severe medical problems such as poisoning or diabetic complications
   - Open chest or abdominal wounds
   - Severe shock

2. Second priority
   - Burns
   - Major multiple fractures
   - Back injuries with or without spinal cord damage

3. Third (lowest) priority
   - Fractures or other minor injuries
   - Obviously mortal wounds
   - Obviously dead

Triage gives the rescuer the chance to keep alive as many persons as possible. Ambulatory victims not needing immediate attention should be taken from the immediate scene as soon as possible to prevent congestion of the area.

Other Actions at the Scene

The person in charge must keep the immediate and surrounding area under as much control as possible. To do this the following are suggested:

- Place warning devices to divert traffic.
- Keep bystanders from crowding or mishandling the victims.
- When necessary, get the assistance of volunteers as in Figure 10.15 (be specific when giving them tasks).
- Reassure relatives; question and inform them away from the victim.

Do not assume the responsibilities of the police or other authorities when they are present. However, do not let their actions interfere with the care of the victims.

Figure 10.15 If required, specifically direct volunteers to assist.

APPENDIX A

Emergency Information Numbers

CHEMICAL TRANSPORTATION EMERGENCY CENTER (CHEMTREC), 1-800-424-9300

CHEMTREC provides a 24-hour hotline service with computer access for inquiries about chemical, radioactive, or explosive emergencies. This free service handles emergencies only, not routine inquires. To use CHEMTREC, gather as much information about the substance in question as possible, ideally from an uncontaminated container label. In addition, CHEMTREC will need as much of the following information as possible: name and organization of caller, call-back number, location of problem, shipper or manufacturer, container type, rail car or truck number, carrier name, consignee, and local conditions.

NATIONAL RESPONSE CENTER (NRC), 1-800-424-8802

The NRC is a hazardous material emergency hotline, operated by the Coast Guard 24 hours a day. The center computer provides information on chemicals or hazardous material shipped by carriers within the United States. The service is rapid, accurate, and free. Information includes emergency action measures for chemical emergencies, pollution effects, and possible neutralizers. NRC needs the same information from the caller as CHEMTREC does.

TRANSPORTATION EMERGENCY ASSISTANCE PROGRAM

This service is a Canadian hazardous material hotline. Its operation is primarily the same as CHEMTREC.

Atlantic provinces and eastern Quebec	819-537-1123
Southwestern Quebec	514-373-8330
Eastern Ontario	613-348-3616
Central Ontario	416-356-8310
Southwestern Ontario	519-339-3711
Northern and western Ontario	705-682-2881
Manitoba, Saskatchewan, Alberta	403-477-8339
British Columbia	604-929-3441

CANUTEC

Another Canadian hotline. Call 613-996-6666.

NATIONAL CAVE RESCUE COMMISSION, 1-618-256-4927

This commission is a free 24-hour emergency information and service center that can be contacted through the United States Air Force Rescue Coordinating Center. This service gives the caller access to cave rescue teams, cave rescue advisors, cave rescue divers and dive teams, cave rescue medical teams, and emergency consultation relating to any type of cave rescue emergency or problem. These services are generally provided at no charge. Callers should ask to talk with a cave rescue coordinator, give the exact location of the emergency, and explain the specific problem.

DIVING ACCIDENT NETWORK

National Hotline Number: 1-918-684-8111

A joint endeavor of the National Oceanic and Atmospheric Administration, the National Institute for Occupational Safety and Health, and the U.S. Department of Energy, the network provides physicians and rescue squads with medical advice for treating diving problems such as gas embolism, bubbles in the bloodstream, and decompression sickness ("the bends").

APPENDIX B

Equipping an Outdoor Search Command Center

These materials should be stocked for an outdoor search command center. Vary quantities to fit the needs of the jurisdiction.

Secretarial supplies (12 search teams)

48	pencils	1	pencil sharpener
12	tablets (8½" × 14")	12	tablets (4" × 6")
12	clipboards (8½" × 14")	1	scissors
1	stapler	1	box staples
12	whistles	1	box paper clips
12	compasses	24	rolls of vinyl tape (assorted colors if possible)
1	industrial first aid kit	13	sets of area maps
1	box thumbtacks	1	hand-wound clock
1	box blackboard chalk	1	resource book
	forms as used by jurisdiction		

Hardware Supplies

1	hammer (16 oz.)	1	set of screw drivers
1	pliers	24	wooden stakes (1' × 2" × 2")
1	garden hose	1	tube of paper cups
1	five-person tent		

APPENDIX C

Synthetic Ropes for Rescue

For John Lyons, technician for the Takoma Park, Maryland, Fire Department, business was as usual on November 21, 1979—until the rope broke.

That day Lyons was conducting a rescue drill for career members on duty. A "victim" was secured in a Stokes basket. The intent was to use an aerial ladder to lift, rotate, then lower the Stokes to the ground. A stretcher bridle was tied using a ¾-inch (20 mm) manila line. Lyons signaled the aerial operator to raise the Stokes (and "victim") so Lyons could instruct the members concerning checking for balance.

Shortly after the Stokes was raised, partly over the roof's edge, the ¾-inch (20 mm) manila rope suddenly snapped about four feet (1.2 m) above the knot. Luckily, Lyons had one hand on the basket and was able to keep it from going over the edge. The other crew members helped him drag the "victim" back to safety.

This incident led Roger A. McGary, chief of the Takoma Park Fire Department, to investigate the reasons for the failure. From rope suppliers he learned two interesting facts that applied directly to his case.

- Because manila rope is made from natural fibers, it deteriorates rapidly over the years and its tensile strength is reduced proportionately.
- If a manila rope ever gets wet, its tensile strength is reduced by half, however well it may be dried afterward. (This applies mostly to the too common practice of wetting a new rope then hanging it to dry with a weight on the free end. A manila rope will also lose half its strength in a year when stored in a humid atmosphere.)

The rope used in the nearly disastrous drill was seven years old and when new had been soaked in water and dried to take out its stiffness.

Chief McGary was appalled, and did some further research. His conclusion at the end was that a synthetic kernmantle rope such as that manufactured by Blue Water or Pigeon Mountain is much better to use for rescue than is manila rope.

Blue Water's largest diameter was ½-inch (13 mm), which many firefighters might think too small for a good hand grip. Chief McGary reports, however, that "from personal experience this is not a major problem." He contends, rightly, that before dismissing the use of synthetic rope it be used and experimented with to determine its advantages and disadvantages. (5/8-inch (16.5 mm) rope has come on the market since McGary's research.)

The information given here should help to dispel much of the confusion concerning synthetic ropes and their usefulness in the fire service. If the matter is still not clear, however, contact your local supplier. A reputable supplier is one who knows the product, stores it properly, and will help you choose the right rope for the job. Synthetic rope costs about the same as manila. Even if it is more expensive remember that being penny wise when buying rope is indeed being pound foolish. There is far too much to lose.

Which Should Be Used?

There is some argument concerning which type of rope is best. Many persons believe that laid nylon rope is better than kernmantle. They contend that inspection for internal damage is easier and surer, that laid nylon has better abrasion resistance over the long run, and that its elasticity is an especial advantage because of the possibility of shock loading during a rescue.

Many other persons prefer the more static kernmantle ropes because of its lesser spin under load. They contend that the tight sheath is more resistant to abrasion because a laid rope, in time, will have had all of its fibers subjected to abrasion. They also assert that elasticity might not be a complete blessing, that there are times when a sudden stretch could cause rescuers to lose control of the victim.

Each department must, in the end, make its own decision. The greatest deciding factors will probably be personal preference and intended use. Other factors that should be considered are local conditions, local practices, and the experience of the personnel who will use the rope.

Synthetic Rope Fibers

A synthetic rope may be made of any one of a number of man-made fibers. The two most common are nylon and Dacron.

The two types of nylon filament used for rope are called *type six* (Allied Chemical and Firestone) and *type six, six* (Monsanto and Dupont). Type six has a melting point of 419-430°F (215-221°C); type six, six has a melting point of 480-500°F (249-260°C). The disadvantages of nylon are that when wet it loses about 5 percent of its strength and energy-absorbing ability (the strength returns when dry) and that it deteriorates if constantly exposed to sunlight. Nylon's greatest advantage is that dry it can absorb about 15,600 foot pounds of force per pound of rope.

Dacron, a Dupont trade name, has a melting point of 482°F (250°C). Dacron's greatest advantage is that sunlight barely affects it and that water does not affect its tensile strength. Dacron's elongation is lower than nylon's but still six times that of manila.

Finishes

Some nylon filament is treated with a chemical called a *finish*. This finish is applied so the filaments will pass through the rope-making machinery more easily and, in some cases, for better abrasion resistance. Some finishes make the rope slick, which could be undesirable during a rescue. The finish will wear off, however, and washing will remove much of it.

Rope Types

There are three basic types of ropes: laid, kernmantle, and plaited.

Laid rope is the type familiar to most people, being the type of construction used for manila rope — bundles of fibers twisted together. Laid nylon rope is elastic because the lays of rope unwind under load.

Kernmantle rope is a core (kern) of filaments surrounded by a sheath (mantle). The core of the rope is 75 percent of the strength of the rope, taking the weight and shock loading to which the rope might be subjected. The way the core is made also determines whether the rope is "dynamic" (elastic) or "static" (not quite so elastic).

Dynamic kernmantle cores are made of several bundles of twisted fibers, similar to a laid rope. The difference is that although the bundles untwist under load to give the rope its stretch, one bundle's untwisting counteracts that of another, so there is not so much spin. (There are also braided cores.)

The fibers of a static kernmantle core are slightly twisted bundles. The only stretch the rope has, then, is that of the nylon: about 17 to 20 percent at failure.

Kernmantle ropes have different kinds of sheaths for different purposes. Some have sheaths that are loosely woven and loose on the core. This type of rope is good when it must be handled often and have knots changed often, because it is flexible and easy to handle. The disadvantages are that the loose weave lets dirt and grit into the core easily, the sheath is less abrasion resistant, and a parted sheath will slip along the core for a great distance.

More tightly woven sheaths are manufactured for static ropes. Such a rope is not as flexible, and knots are harder to tie in it, but the sheath is more grit and abrasion resistant and will not slip as far if parted.

Care of Synthetic Rope

Store it properly. Keep the rope neatly coiled and out of strong sunlight.

Protect it from harmful substances. Synthetic ropes, like manila ropes, are harmed by a number of substances, among which are benzine, phenol, carbon tetrachloride, formaldehyde, and bleach. Nylon is especially easily damaged by acid. Keep it away from all types of batteries. Although petroleum products are said to be harmless to nylon, they should be kept off nylon rope or at least cleaned from the rope as soon as possible.

Do not walk or stand on it. Walking or standing on a rope grinds dirt and grit into it.

Protect it from abrasion. Use hose rollers or some other item (for example, a folded turnout coat) when a rope must pass over an edge. A rope passing over an edge has only about 50 percent of its strength at that point.

Use it for its intended purpose. A rescue rope intended for hoisting or lowering people must be used only for that purpose. For example, using a rope that has towed a vehicle could very well end in disaster.

Keep it clean. Some persons use a commercial rope washer; others use a regular clothes washer and dryer. In the latter case, be careful that the agitator does not abrade the rope. Do not use hot water or bleach. Use a mild soap. A washed and dried rope will shrink by about 2 percent. (Remember not to dry the rope in direct sunlight.)

Inspect it after each use. Whatever kind of rope is used, run every foot of it through your hands before using it again. Only in this way can differences in flexibility, diameter, or a lumpiness be detected—sure signs of something wrong. If there is any suspicion of damage, immediately cut the rope at that point. Surface fuzz is natural after a rope has been used, and a *small* amount may help guard the rope against further abrasion damage. A badly frayed or abraded rope, however, should be suspect. Exterior condition is often a clue to interior damage.

Keep nylon from running over nylon. The heat of friction can cause failure.

Buying Synthetic Rope

As was mentioned earlier, buying a rope for rescue is a serious business and must be done with much thought. A department must buy the best rope the budget will allow, and buy it from a reputable and knowledgeable supplier (which is most likely not the local hardware or farm supply store.)

Get all the information possible from the supplier and manufacturer. "Static tensile strength," for example, is useful to an extent, but a rope will seldom fail from just stretching at a rescue. Ropes fail during rescues because of previous use or abuse, abrasion, knotting, and other use-caused reasons. Find out the failure characteristics of the rope when knotted, among other things.

APPENDIX D

Rescue From Low-Head Dams and Other Hydraulics

The "drowning machine" created by the reversing currents at the base of a low-head dam or other depression in fast moving water such as a waterfall or large rock requires special rescue procedures.

For rescues where the waterway is not extremely wide, the Dayton, Ohio, Fire Department has used the fire hose float to rescue persons from low-head dams. This concept uses available fire service equipment to construct a rescue device for immediate action. An air cylinder or carbon dioxide fire extinguisher can be used to inflate a section of 2½ or 3-inch fire hose to make the float. Dayton uses the following components for its float.

- 2½-inch (65 mm) fire hose.
- One 2½-inch (65 mm) female cap.
- One 2½-inch (65 mm) male cap with an 18-inch (450 mm), ½-inch (13 mm) flexible hose. There is a relief valve, set at 120 psi (827 kPa) at the male cap, and a shutoff. The ½-inch (13 mm) hose has a quick-connect coupling for a 30-minute (45 ft.^{3} 1,270 L) SCBA cylinder.
- Rope to help guide hose.
- Hooks (to snag unconscious or dead victims).
- Hose straps (to secure hooks to hose).
- Bright orange foam float (so personnel can see end of hose easier).

A 50-foot (15 m) section of hose can be filled to a recommended pressure of 100 psi (700 kPa) in about six seconds from a 45 ft.^{3} (1,270 L) air cylinder. Each hose section will lower the cylinder's pressure by about 500 to 600 psi (3,440 to 4,000 kPa). A single cylinder should fill five sections of 2½-inch (65 mm) hose.

The Dayton procedure is:

Step 1: Get the necessary equipment to the emergency scene.

Step 2: Connect together the number of hose sections needed to reach the victim. Tighten the couplings to prevent air leaks.

Step 3: Lay the hose in U-shaped folds along the bank at the hydraulic or reversal which is the name for the reversing of the current flow that causes the rolling circular motion at the downstream face of the dam. Place the couplings at the edge of the water (Figure D.1).

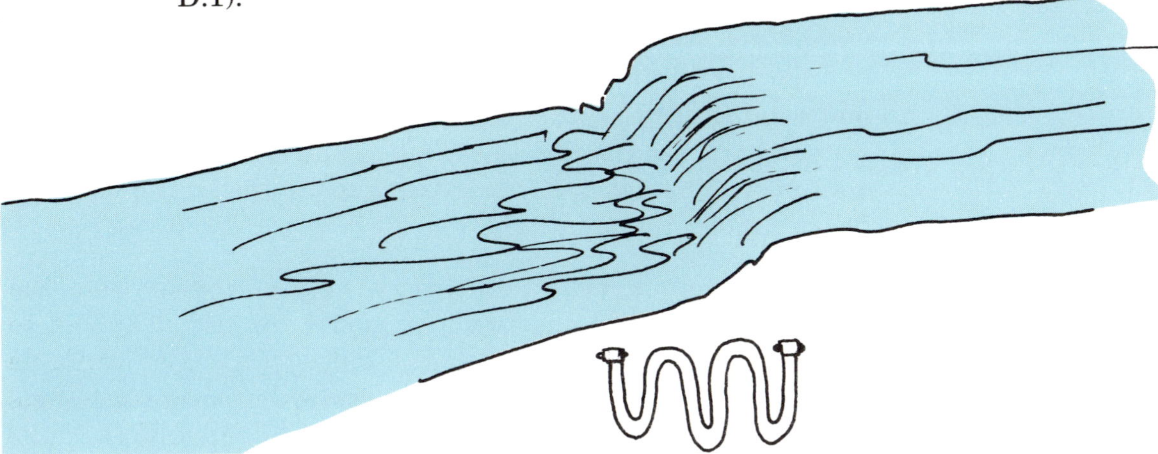

Figure D.1 Lay the fire hose in U-shaped folds on the bank at the low-head dam hydraulic. Place both couplings at the water's edge.

Step 4: Charge the fire hose with air or CO_2 to 100 psi (700 kPa).

Step 5: Extend the capped male end toward the victim. This end should not enter the water until it is about ten feet (3 m) from the bank. This prevents the current from diverting the hose downstream away from the victim. If necessary, use a rope to keep the end out of the water for the 10-foot (3 m) distance (Figure D.2). On wider waterways it may be necessary to pull the male end from the opposite bank if rescuers can reach the other side. A rope gun may be of assistance in positioning lines on the opposite bank.

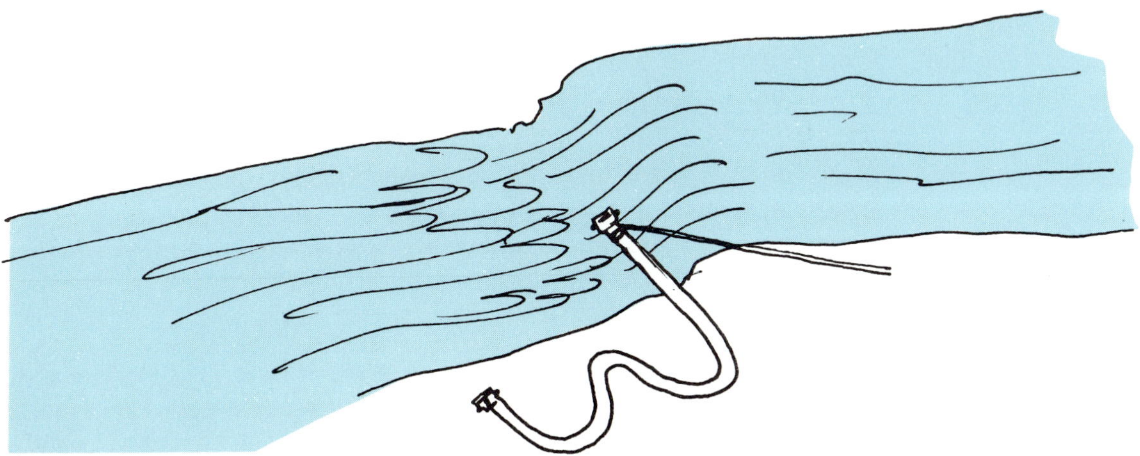

Figure D.2 Use a rope, if necessary, to keep the end of the hose from entering the water until the end is at least ten feet (3 m) from shore.

Step 6: Continue to feed the hose across the hydraulic to the victim.

Step 7: Have the victim hold onto the hose.

Step 8: Pull the victim to safety.

On large fast-flowing water it may be necessary to use an inflatable rescue craft. This should be secured with a four point tie in order to permit movement in all directions. Accessibility may make it difficult to secure the bow and stern to both sides of the waterway. For maximum safety an unmanned trial run with the craft is recommended. Rescuers assigned to the craft must wear full flotation equipment.

GLOSSARY

A-FRAME An inverted-V frame of metal or wood from which is suspended a pulley or block and tackle through which rope or cable passes for lifting objects.

AIR BAG An inflatable bag, often made of synthetic rubber, used to lift or stabilize heavy objects. Also, a large inflatable bag into which persons can leap to escape danger.

ASCENDER A mechanical contrivance, used when climbing rope, that allows upward movement but not downward movement.

BREAST TIMBER A strut that holds a horizontal compression load, keeping sheeting in place for shoring.

CANUTEC A Canadian hazardous material hotline.

CARABINER A D-shaped metal device used in connection with ascending or descending a rope.

CARDIAC NEUROSIS Stress-caused reaction resembling a heart attack in a healthy heart.

CHEMTREC (Chemical Transportation Emergency Center) A 24-hour emergency hotline that can be used to get information concerning hazardous materials.

COMPOUND TACKLE Two or more blocks reeved with more than one rope.

CONVERSION HYSTERIA The unfounded belief that a part of the body has ceased normal function.

CRIBBING A column of overlapped timbers used to support unstable weight.

DEAD SHORE A shore that holds a vertical load; for example, an unstable floor.

DIVING ACCIDENT NETWORK A hotline that can advise physicians and rescue personnel concerning diving mishaps.

DOSIMETER A small, nuclear-radiation-detection device that registers the total amount of radiation to which it has been exposed.

DROWNING MACHINE A colloquial term for the convection currents resulting when water floods over a low-head dam.

DYNAMIC A rope that will stretch farther than will a static rope (see *Static*).

EXPOSURE BAG Special neoprene bag into which a person is placed for field treatment of hypothermia.

FALL LINE The line on which the pull is exerted when using a block and tackle.

FIGURE EIGHT A forged metal device in the shape of an eight; used to help control the speed of a person descending a rope.

FIRE HOSE FLOAT A water-rescue flotation device made of inflated fire hose.

FLYING SHORE A shore for vertical surfaces, such as a wall, that is braced against another vertical surface.

GIBBS CAM An ascender for rope climbing.

GIN POLE A guyed pole, angled over a load to be lifted, near the top of which is attached a pulley or block and tackle for lifting the load.

HEEL The foot plate against which a breast timber is footed.

HOLDFAST A constructed anchor for a guy line.

HYPERVENTILATION Rapid breathing that overoxygenates the blood.

HYPOTHERMIA The condition in which the human body loses heat to its environment and which can lead to death.

JUMAR An ascender for rope climbing.

KERNMANTLE Also called "core and sheath." A rope constructed with a core of straight fibers around which is a sheath of woven fibers.

LADDER FLOAT An inflated tire or tire inner tube fastened to a ladder; used for rescuing persons from water.

LADDER GIN POLE A gin pole in which the load-supporting member is a straight ladder.

LAID A type of rope constructed of fiber bundles twisted together.

LEADING BLOCK A block that changes the direction of pull without affecting the mechanical advantage of a block-and-tackle arrangement.

LINEMAN'S GLOVES Special gloves through which electricity cannot be conducted.

MAMMALIAN DIVING REFLEX The autonomous physiologic reaction to immersion in cold water, in which the blood and oxygen supply is shunted to the brain to keep the animal alive although outward appearance may suggest death.

MINE RESCUE DRILL Special drill, operated by the Mine Emergency Division of the U.S. Mine Safety and Health Administration, that can drill a 24-inch-diameter (0.6 m) shaft through 50 feet (15 m) of solid limestone in one day.

MOUSING Wraps of cord or wire used to keep a rope or cable in a hook used with lifting equipment.

NATIONAL CAVE RESCUE COMMISSION A division of the National Speleological Society, specializing in rescuing persons from caves.

NIELLS-ROBERTSON A stretcher that immobilizes a victim, prevents further spinal damage, and protects the head.

PACKAGING Readying a victim for transport.

PICKET A stake driven into the ground to anchor a guy line.

POMPIER BELT Lifebelt.

PRUSSIK KNOT A special knot used to assist a person climb a rope.

RAKER The angled timber bearing the load exerted on a raking shore.

RAKING SHORE A shore footed on a horizontal surface and used to brace a vertical surface.

RATE METER A nuclear-radiation-detection device.

REEVING Threading rope or cable through a block.

RIGGING Ropes or cables used with lifting or pulling devices such as block and tackle.

RUNNING BLOCK The block that moves in a block-and-tackle arrangement.

SHORING BLOCK A shim for a jack.

SIMPLE TACKLE One or more blocks reeved with a single rope.

SOLE PLATE The member against which the vertical load of a shore is ultimately exerted.

SPACER A length of timber that keeps a breast timber from shifting vertically.

STANDING BLOCK The stationary block of a block-and-tackle arrangement.

STATIC A rope that will stretch a relatively short distance under load.

STRAINING PIECE A length of timber that keeps pressure on the breast timbers of a flying shore.

STRUT In a shore, any member that holds either a vertical or horizontal compression load.

SURVEY METER A nuclear-radiation-detection instrument.

SWISS SEAT A harness that keeps a person's center of gravity near normal while rappelling.

TRENCH JACK Any jack used to keep sheeting or wales apart for the insertion of a breast timber. Also, a jack that is used as a breast timber.

VERTICAL SHORE A dead shore.

WALE A timber placed against shoring sheeting planks that, when a breast timber is in place, keeps the sheeting planks in place.

WALL PLATE In a raking shore or a flying shore, the continuous sheeting member placed immediately against the vertical surface shored.

WARM-GAS INHALATOR A device that warms air for a hypothermia victim to breathe.

INDEX

IFSTA MANUALS

SERVICE ORIENTATION & INDOCTRINATION
ry, traditions, and organization of the fire service; operation of the fire depart- and responsibilities and duties of firefighters; fire department companies and functions; glossary of fire service terms.

SERVICE FIRST AID PRACTICES
explanations of the nervous, skeletal, muscular, abdominal, digestive, and ourinary systems; injuries and treatment relating to each system; bleeding ol and bandaging; artificial respiration, cardiopulmonary resuscitation (CPR), ., poisoning, and emergencies caused by heat and cold; fractures, sprains, islocations; emergency childbirth; short-distance transfer of patients; ambu- s; conducting a primary and secondary survey.

ENTIALS OF FIRE FIGHTING
manual meets the objectives set forth in levels I and II of NFPA *Fire Fighter ssional Qualifications, 1981.* Included in the manual are the basics of: fire be- r, extinguishers, ropes and knots, self-contained breathing apparatus, lad- forcible entry, rescue, water supply, fire streams, hose, ventilation, salvage overhaul, fire cause determination, fire suppression techniques, communica- sprinkler systems, and fire inspection..

A'S 500 COMPETENCIES FOR FIREFIGHTER CERTIFICATION
nanual identifies the competencies that must be achieved for certification as a hter for levels one and two. The text also identifies what the instructor needs e the student, NFPA standards, and has space to record the student's score, standards, and the instructor's initials.

SERVICE GROUND LADDER PRACTICES
us terms applied to ladders; types, construction, maintenance, and testing of ervice ground ladders; detailed information on handling ground ladders and al tasks related to them.

HOSE PRACTICES
struction, care, and testing of hose and various fire hose accessories; prepara- and manipulation of hose for rolls, folds, connections, carries, drags, and spe- perations; loads and layouts for fire hose.

VAGE AND OVERHAUL PRACTICES
ning and preparing for salvage operations, care and preparation of equipment, ods of spreading and folding salvage covers, most effective way to handle r runoff, value of proper overhaul and equipment needed, and recognizing and erving arson evidence.

CIBLE ENTRY, ROPE AND PORTABLE INGUISHER PRACTICES
s of forcible entry tools and general building construction; use of tools in open- loors, windows, roofs, floors, walls, partitions and ceilings; types, uses, and of ropes, knots, and portable fire extinguishers.

F-CONTAINED BREATHING APPARATUS
manual is the most comprehensive self-contained breathing apparatus text able. Beginning with the history of breathing apparatus and the reasons they eeded, to how to use them, including maintenance and care, the firefighter is n step by step with the aid of programmed-learning questions and answers ighout to complete knowledge of the subject. The donning, operation, and care types of breathing apparatus are covered in depth, as are training in SCBA breathing-air purification, and recharging cylinders. There are also special oters on emergency escape procedure and interior search and rescue.

E VENTILATION PRACTICES
ctives and advantages of ventilation; requirements for burning, flammable liq- haracteristics and products of combustion; phases of burning, backdrafts, and transmission of heat; construction features to be considered; the ventilation ess including evaluating and size up is discussed in length.

E SERVICE RESCUE PRACTICES
TA's new *Rescue* has been enlarged and brought up to date. Sections include r and ice rescue, trenching, cave rescue, rigging, search-and-rescue tech- es for inside structures and outside, and taking command at an incident. Also ded are vehicle extrication and a complete section on rescue tools. The book rs all the information called for by the rescue sections of NFPA 1001 for Fire ter I, II, and III, and is profusely illustrated.

THE FIRE DEPARTMENT COMPANY OFFICER
This manual focuses on the basic principles of fire department organization, working relationships, and personnel management. For the firefighter aspiring to become a company officer and the company officer who wishes to improve management skills this manual will be invaluable. This manual will help individuals develop and improve the necessary traits to effectively manage the fire company.

FIRE CAUSE DETERMINATION
Covers need for determination, finding origin and cause, documenting evidence, interviewing witnesses, courtroom demeanor, and more. Ideal text for company officers, firefighters, inspectors, investigators, insurance, and industrial personnel.

PRIVATE FIRE PROTECTION & DETECTION
Automatic sprinkler systems, special extinguishing systems, standpipes, detection and alarm systems. Includes how to test sprinkler systems for the firefighter to meet NFPA 1001.

INDUSTRIAL FIRE PROTECTION
Past loss experience shows, without a doubt, that devastating fires in industrial plants do occur. They occur at a rate of 145 industrial fires every day. *Industrial Fire Protection* is the single source document designed for training and managing industrial fire brigades.

This text is a must for all industrial sites, large and small, to meet the requirements of the Occupational Safety and Health Administration's (OSHA) regulation 29 CFR part 1910, Subpart L, concerning incipient industrial fire fighting.

FIRE SERVICE INSTRUCTOR
Characteristics of good instructor; determining training requirements and what to teach; types, principles, and procedures of teaching and learning; training aids and devices; conference leadership.

PUBLIC FIRE EDUCATION
A valuable contribution to your community's fire safety. Includes public fire education planning, target audiences, seasonal fire problems, smoke detectors, working with the media, burn injuries, and resource exchange.

FIRE PREVENTION AND INSPECTION PRACTICES
Fire prevention bureau and inspecting agencies; fire hazards and causes; prevention and inspection techniques; building construction, occupancy, and fire load; special-purpose inspections; inspection forms and checklists along with reference sources; maps and symbols; records and reports.

WATER SUPPLIES FOR FIRE PROTECTION
Importance, basic components, adequacy, reliability, and carrying capacity of water systems; specifications, installation, maintenance, and distribution of fire hydrants; flow test and control valves; sprinkler and standpipe systems.

FIRE APPARATUS PRACTICES
Various types of fire apparatus classified by functions; driving and operating apparatus including pumpers, aerial ladders, and elevating platforms; maintenance and testing of apparatus.

FIRE STREAM PRACTICES
Characteristics, requirements, and principles of fire streams; developing, computing, and applying various types of streams to operational situations; formulas for application of hydraulics; actions and reactions created by applying streams under different circumstances.

FIRE PROTECTION ADMINISTRATION
A reprint of the Illinois Department of Commerce and Community Affairs publication. A manual for trustees, municipal officials, and fire chiefs of fire districts and small communities. Subjects covered include officials' duties and responsibilities, organization and management, personnel management and training, budgeting and finance, annexation and disconnection.

FIREFIGHTER SAFETY
Basic concepts and philosophy of accident prevention; essentials of a safety program and training for safety; station house facility safety; hazards enroute and at the emergency scene; personal protective equipment; special hazards, including chemicals, electricity, and radioactive materials; inspection safety; health considerations.

FIRE PROBLEMS IN HIGH-RISE BUILDINGS
Locating, confining, and extinguishing fires; heat, smoke, fire gases, and life hazards; exposures, water supplies and communications; pre-fire planning, ventilation, salvage and overhaul; smokeproof stairways and problems of building design and maintenance; tactical checklist.

PHOTOGRAPHY FOR THE FIRE SERVICE
Camera components and operations, films — their advantages and differences, fundamental principles in taking a good picture, processing the negative and print, controlling light for optimum results, accountability and storage of photographic materials; fire scene photography, using photographs as training aids, enhancing fire prevention and public relations through photography, how the investigator can use photographs, glossary of photographic terms.

AIRCRAFT FIRE PROTECTION AND RESCUE PROCEDURES
Aircraft types, engines, and systems, conventional and specialized fire fighting apparatus, tools, clothing, extinguishing agents, dangerous materials, communications, pre-fire planning, and airfield operations.

GROUND COVER FIRE FIGHTING PRACTICES
Ground cover fire apparatus, equipment, extinguishing agents, and fireground safety; organization and planning for ground cover fire; authority, jurisdiction, and mutual aid, techniques and procedures used for combating ground cover fire.

FIRE SERVICE PRACTICES FOR VOLUNTEER AND SMALL COMMUNITY FIRE DEPARTMENTS
A general overview of material covered in detail in *Forcible Entry, Ladders, Hose, Salvage and Overhaul, Fire Streams, Apparatus, Ventilation, Rescue, Inspection,* and *Self-Contained Breathing Apparatus,* and *Public Fire Education.*

INSTRUCTOR GUIDE SETS
Available for *Forcible Entry, Ladder, Hose, Salvage and Overhaul, Fire Streams, Apparatus, Ventilation, Rescue, First Aid, Inspection, Aircraft, and for the slide program Fire Department Support of Automatic Sprinkler Systems.* Basic lesson plan, tips for instructor, references. *Essentials* has NFPA Standard 1001 references and pertinent proficiency tests.

TRANSPARENCIES
Multicolored overhead transparencies to augment *Essentials of Fire Fighting* are now available. Since costs and availability vary with different chapters, contact IFSTA Headquarters for details. Units available:

Fire Behavior; Portable Extinguishers; Ropes and Knots; Hose Tools and Appliances; Handling Hose; Handling Ground Ladders; Ventilation; Fire Streams; Ladder Carries and Raises; Forcible Entry; Salvage and Overhaul; Prevention and Identification; Ground Cover Fires; Communications; Water Supply; Automatic Sprinkler Systems; Rescue; Protective Breathing Apparatus.

SLIDES
2-inch by 2-inch slides that can be used in any 35 mm slide projector; supplem to respective manuals and sprinkler guide sets.

Ladders
Sprinklers
Module 1: Introduction to Automatic Sprinkler Protection
Module 2: Types of Sprinkler Systems
Module 3: Maintenance and Inspection of Sprinkler Systems
Module 4: Components of Water Supply Systems
Module 5: Testing and Analysis of Water Supply Systems
Module 6: Factors Affecting the Adequacy of Sprinkler and Water Syster
Smoke Detectors Can Save Your Life
Matches Aren't For Children
Public Relations for the Fire Service
Public Fire Education Specialist (Slide/Tape)
Salvage*

*The complete package consists of the slides, instructor's manual, and instruc guide sets.

MANUAL HOLDER
The fast, efficient way to organize your IFSTA manuals. These attractive he duty vinyl holders have specially designed side panels that allow easy access t of your IFSTA manuals. Manual holders stand unsupported and will hold up to e manuals.

GUIDE SHEET BINDERS
Free with purchase of complete guide set. Binders also available separately.

WATER FLOW TEST SUMMARY SHEETS
50 summary sheets and instructions on how to use; logarithmic scale to simplify process of determining the available water in an area.

PERSONNEL RECORD FOLDERS
Personnel record folders should be used by the training officer for each membe the department. Such data as IFSTA training, technical training (seminars), college work can be recorded in the file, along with other valuable information. ter size or legal size.

o: Date ____________

Customer Number

ization (FOR INVOICE TO ONLY) Phone

ss

State Zip

Send to
Fire Protection Publications
Oklahoma State University
Stillwater, Oklahoma 74078
(405) 624-5723
Or Contact Your Local Distributor

ORDER FORM

IFSTA MANUALS

WRITE THE NUMBER OF COPIES OF EACH MANUAL NEXT TO ITS TITLE.

	No. Of Each
ination	______
d	______
tials	______
ompetencies for tials	______
rs	______
	______
ge and Overhaul	______
le Entry	______
ontained Breathing aratus	______
ation	______
le	______
any Officer	______

	No. of Each
Fire Cause Determination	______
Private Fire Protection	______
Industrial Fire Protection	______
Instructor	______
Public Fire Education	______
Fire Prevention/ Inspection	______
Water Supplies	______
Fire Streams	______
Apparatus Practices	______
Fire Protection Administration	______

	No. of Each
Safety	______
Aircraft	______
Photography	______
High-Rise	______
Volunteer	______
Ground Cover	______
Manual Holder	______
SLIDES	
Ladder	______
Salvage	______
Public Fire Education Specialists	______

	No. of Each
Smoke Detectors Can Save Your Life	______
Matches Aren't For Children	______
Public Relations for the Fire Service	______

TRANSPARENCIES AND SLIDES
Multicolored overhead transparencies to augment each chapter of *Essentials of Fire Fighting* are now available. Also available are slide programs for each of the major fire pump manufacturers and slides for sprinkler systems. Since costs and availability vary with different sets, contact Fire Protection Publications for details.

OTHER MANUALS AND MATERIALS MAY BE ORDERED BELOW:

JANTITY	TITLE	LIST PRICE	TOTAL

All Foreign Orders must be prepaid in U.S. currency and include 20% shipping and handling charges.

Free Subscription to Speaking of Fire ☐

ain postage and prices from current IFSTA Catalog or they will be inserted by tomer Services.

e: Payment with your order saves you postage and handling charges n ordering from Fire Protection Publications.

ment Enclosed ☐ Bill Me Later ☐

w 4 to 6 weeks for delivery.

SUBTOTAL	$ ____________
Discount, if applicable	$ ____________
Postage and Handling, if applicable	$ ____________
TOTAL	$ ____________

FOR ORDERS
TOLL FREE NUMBER — 800-654-4055

Oklahoma, Hawaii, and Alaska call collect.